UNIVERSITY MATHEMATICAL TEXTS

EDITORS/ALAN JEFFREY AND IAIN T. ADAMSO

INTRODUCTION TO METRIC SPACES

RECENT UNIVERSITY MATHEMATICAL TEXTS

An Introduction to Modern Mathematics/Albert Monjallon
Quantum Mechanics/R. A. Newing and J. Cunningham
Probability/J. R. Gray
Series Expansions for Mathematical Physicists/H. Meschkowski
Calculus of Variations/J. C. Clegg
Elementary Rings and Modules/Iain T. Adamson
An Introduction to the Theory of Statistics/R. L. Plackett
Mathematical Analysis/G. H Fullerton

C. G. C. PITTS
LECTURER IN MATHEMATICS
UNIVERSITY OF EAST ANGLIA

INTRODUCTION TO METRIC SPACES

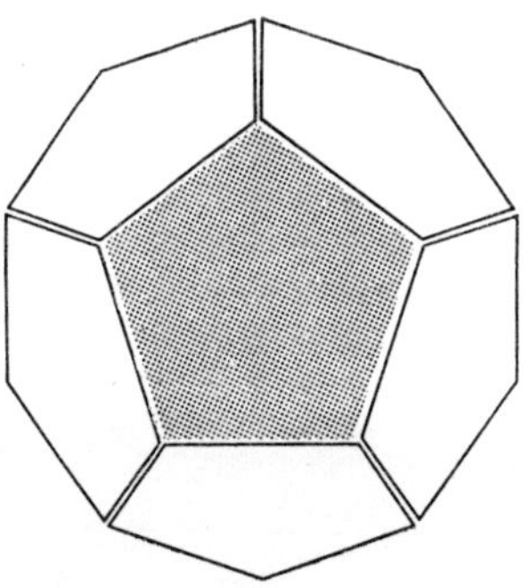

OLIVER & BOYD EDINBURGH

OLIVER AND BOYD
Tweeddale Court
14 High Street
Edinburgh EH1 1YL

A Division of Longman Group Limited

UMT 42
ISBN 0 05 002453 1
First published 1972

Printed in Great Britain by
Bell and Bain Ltd., Glasgow

PREFACE

The importance of functional analysis to many branches of mathematics has been realised since the 1930's; however it is only comparatively recently that the subject has formed any significant part of undergraduate mathematics courses. Metric spaces, as the simplest system to be studied within the field of functional analysis, form a natural and necessary introductory topic. Moreover they arise as a result of the abstraction of the distance properties of the real line (or of points in real m-dimensional space). The topic therefore unifies much of the theory of elementary analysis and bridges the gap between classical and functional analysis. In consequence this book has been written in a style suitable for a student who has completed elementary analysis courses up to and including uniform convergence. (It is the author's belief that most students are best introduced to uniform convergence in the classical way rather than via the supremum metric for the set of continuous, or bounded, functions.) Throughout the text, the introduction of each concept has been carefully motivated by showing how it arises naturally as the abstraction of a corresponding idea in elementary analysis.

There are a few details concerning layout, etc. Chapter 1 is a catalogue of results needed for the main part of the text which starts in Chapter 2. All of Chapter 1, with the possible exception of §1.8 (concerning Minkowski's inequality), is assumed to be familiar to the reader; accordingly there are no exercises in this chapter. There are, however, over 200 exercises distributed over the remaining chapters; most of these require only routine manipulation of the theory, but some give further results. Within the actual text there are occasional gaps in the arguments which the reader is invited to fill for himself; these gaps are always indicated, and require only straightforward steps.

There is one further detail. Except in Chapter 1, any major result that is stated and which is assumed to be known to the reader is called a 'proposition'; this distinguishes such results from those proved in the text, and which in the usual way are designated as theorems, lemmas, or corollaries. Such propositions are stated from time to time to provide an introduction to, and a motivation of, the more abstract ideas described for general metric spaces.

I am indebted to many of my colleagues for comments on the earlier versions of this book. Undoubtedly my greatest debt, which I gratefully acknowledge, is to Dr. L. E. Clarke who, as well as discussing with me various aspects of this book, has also read in detail several of the earlier versions and has worked nearly all the exercises; he has pointed out a number of errors and obscurities, and suggested many improvements. Needless to say, however, the final responsibility of any remaining errors remains with me; I should much appreciate having these reported to me. Lastly I am pleased to acknowledge the assistance of my wife in typing the final draft of this book.

C. G. C. Pitts.

School of Mathematics and Physics,
University of East Anglia,
Norwich.

CONTENTS

1 : PRELIMINARIES

In this chapter there are listed some terminology and results which will be required subsequently; the reader is assumed to be familiar with these ideas so they are discussed briefly and the proofs of the results are not, in general, given. Thus the contents of this chapter are to be regarded as a catalogue of material which will be required at various places in the remainder of the text, and (with the exception of the last section) are not to be regarded as a rigorous development of these topics.

The topics to be reviewed are firstly sets, mappings, equivalence relations and countability; these are all treated more fully by Simmons (1963) §§1–7. Secondly, in §§1.5, 1.6 there are listed a number of elementary results concerning the real number system and real functions; for a full discussion of this material see any first course in analysis (for example Apostol (1957)). In §1.7 we discuss the negation of the quantifiers 'there exists' and 'for all'. Lastly, in §1.8 some inequalities are given; since the reader may not have met these before their proofs are given in full.

1.1 Sets

A *set* is a collection of objects; the latter are referred to as the *elements*, *members*, or *points* of the set. If X is the set of all elements x such that some proposition $P(x)$ is true, we shall write $X = \{x : P(x)\}$. If x is a member of a set X, we write $x \in X$; if x is not a member of X, we write $x \notin X$.

The following notation will be used throughout this text. N is the set of all natural numbers (1, 2, ...), Z is the set of all integers, Q is the set of all rationals, R is the set of all real numbers, and C is the set of all complex numbers.

The empty set will be denoted by $\varnothing$; it is a subset of every set.

Two sets are said to be *equal* if and only if every element of each set is also an element of the other. If A, B are two sets such that all the elements of A are also elements of B, then we say that A is a *subset* of B and we write $A \subseteq B$; the symbol $\subset$ is reserved for strict inclusion. If $\varnothing \subset A \subset B$ then A is said to be a *proper* subset of B. If A, B are any two sets, their union is denoted by $A \cup B$ and their intersection by $A \cap B$. If $A \cap B = \varnothing$ then the sets A, B are said to be *disjoint*.

A finite set is one containing only a finite number of elements, and an infinite set is one containing an infinite number of elements.

For linguistic convenience we shall also employ the terms *collection* and *class* instead of set; in particular these will be used to avoid talking of 'a set of sets' and to say alternatively 'a class of sets' or 'a collection of sets'. Thus

$$\{X_n : n = 1, 2, \ldots\}, \text{ that is } \{X_n : n \in \mathsf{N}\} \tag{1.1.1}$$

is an infinite class of sets.

More generally, if Λ is any set and for each λ in Λ a (unique) set X_λ is defined, then $\{X_\lambda : \lambda \in \Lambda\}$ is a class of sets; Λ is called an *index* set for this class. Thus in (1.1.1) N is an index set.

The identities

$$A \cap (B \cup C) = (A \cap B) \cup (A \cap C),$$

$$A \cup (B \cap C) = (A \cup B) \cap (A \cup C),$$

are known as the distributive rules for set union and intersection. More generally, if $\{B_\lambda : \lambda \in \Lambda\}$ is any collection of sets, we have

$$A \cap \left(\bigcup_{\lambda \in \Lambda} B_\lambda\right) = \bigcup_{\lambda \in \Lambda} (A \cap B_\lambda), \tag{1.1.2}$$

$$A \cup \left(\bigcap_{\lambda \in \Lambda} B_\lambda\right) = \bigcap_{\lambda \in \Lambda} (A \cup B_\lambda). \tag{1.1.3}$$

If X, Y are two sets, then the set of all elements in X which are not in Y is called the *complement* of Y with respect to X, and is denoted by $X - Y$; thus

$$X - Y = \{x : x \in X, x \notin Y\}.$$

(Some authors write $X \backslash Y$ instead of $X - Y$.)

Let $\{A_\lambda : \lambda \in \Lambda\}$ be a collection of subsets of a set X; then

$$X - \bigcup_{\lambda \in \Lambda} A_\lambda = \bigcap_{\lambda \in \Lambda} (X - A_\lambda),$$

and

$$X - \bigcap_{\lambda \in \Lambda} A_\lambda = \bigcup_{\lambda \in \Lambda} (X - A_\lambda).$$

These relations are known as De Morgan's rules.

If X_1, X_2 are two sets, then the set of all ordered pairs (x_1, x_2) where $x_1 \in X_1, x_2 \in X_2$ is called the *Cartesian product* of X_1 and X_2, and is denoted by $X_1 \times X_2$; thus

$$X_1 \times X_2 = \{(x_1, x_2) : x_1 \in X_1, x_2 \in X_2\}.$$

Likewise we may define the Cartesian product $X_1 \times \ldots \times X_n$ of the sets $X_1, \ldots, X_n$; if $X_1 = \ldots = X_n = X$, say, then we write $X_1 \times \ldots \times X_n = X^n$.

More generally still we may define the Cartesian product of any infinite sequence of sets; thus

$$X_1 \times X_2 \times \ldots = \{(x_1, x_2, \ldots) : x_i \in X_i\}.$$

The elements of $X_1 \times X_2 \times \ldots$ are infinite sequences (the elements of $X_1 \times \ldots \times X_n$ may be regarded as finite sequences).

Finally we mention the word *space*, which is often used instead of set, collection etc. to specify the universe of elements in which we happen to be interested. The space will usually have some structure which will be indicated by an adjective describing it (for example, vector space or metric space).

1.2 Functions

Let X, X' be non-empty sets; if a rule is given, by which to each x in X there is assigned a unique corresponding x' in X', then the rule is said to be a *function*, or *mapping*. If this rule is denoted by f, then we say that f is a *mapping of X into X'* and write $f: X \to X'$.

This description of what is meant by a function gives us the necessary criteria to decide whether or not a given rule is a function; however it does not strictly define the concept of a function since it merely replaces the word function by the (undefined) word rule. This criticism can be countered as follows.

Let X, X' be non-empty sets; any subset of $X \times X'$ such that for each x in X there exists one and only one element (x, x') in the subset is said to be a *function*, or *mapping*, on X. If the subset is denoted by f, then we say that f is a *mapping of X into X'* and write $f: X \to X'$.

For our purposes either definition of a function is acceptable and the reader should adopt whichever he prefers.

We shall use the terms function and mapping interchangeably.

If $f: X \to X'$ is a mapping and if $x \in X$, the corresponding x' (known as the image of x under f) in X' is denoted by $f(x)$ so $x' = f(x)$.

Occasionally it will be convenient to use an alternative notation. We shall then denote a function $f: X \to X'$ by $x \mapsto f(x)$, $x \in X$; for example 'the function $f: \mathsf{R} \to \mathsf{R}$ defined by $f(x) = \sin x$, $x \in \mathsf{R}$' can be written more briefly as 'the function $x \mapsto \sin x$, $x \in \mathsf{R}$'. (This notation does not give the set X', but this information is usually self-evident.)

Some set-theoretic terminology and results concerning a mapping $f: X \to X'$ will now be described.

If A is any subset of X, then the set of images of all elements of A, under the mapping f, is denoted by $f(A)$; thus

$$f(A) = \{f(x): x \in A\}. \tag{1.2.1}$$

X is called the *domain*, and $f(X)$ is called the *range*, of f.

If $f(X) = X'$, then f is said to be *surjective* (or f is said to map X *onto* X'). If every pair of distinct elements x, y of X map into distinct elements $f(x), f(y)$ of X', f is said to be *injective* (or *one-to-one*). If f is injective and surjective, then it is said to be *bijective*.

Note that $f: X \to X'$ possesses an inverse (defined on X') if and only if f is bijective. This inverse is denoted by $f^{-1}: X' \to X$.

Given a mapping $f: X \to X'$, if $Y \subseteq X$ and if $g: Y \to X'$ is a mapping such that $g(y) = f(y)$ for all y in Y, then g is called a *restriction* of f and f is called an *extension* of g; this restriction g of f is denoted by f_Y and is called the restriction of f to Y.

If $f: X \to X'$ and $g: X' \to X''$ are two mappings, then the mapping $h: X \to X''$ defined by $h(x) = g\{f(x)\}$ for all x in X is called the *composition* of f and g; it is denoted by $h = g \circ f$.

If X is a non-empty set, the function $f: X \to X$ defined by $f(x) = x$ for all x in X is called the *identity* function on X; often the symbol i is used instead of f for this function.

Again let $f: X \to X'$ be an arbitrary function and let $A' \subseteq X'$; then the set of all elements of X which are mapped into A' by f is called the *inverse image* of A' and is denoted by $f^{-1}(A')$, that is

$$f^{-1}(A') = \{x: x \in X, f(x) \in A'\}. \tag{1.2.2}$$

It should be noted that in order to define $f^{-1}(A')$ it is *not* necessary that f should possess an inverse. Thus the notation f^{-1} has been used to represent two distinct ideas; however no ambiguity can arise.

Since the mapping f is not necessarily surjective, there may be elements in A' which are not images of any point in X; however this does not invalidate the above definition of $f^{-1}(A')$. Thus, in particular, if $A' \cap f(X) = \varnothing$ then $f^{-1}(A') = \varnothing$.

There are a number of relations concerning the sets $f(A), f^{-1}(A')$ defined in (1.2.1), (1.2.2).

If $A \subseteq B \subseteq X$ then

$$f(A) \subseteq f(B). \tag{1.2.3}$$

If $A_1, A_2 \subseteq X$ then

$$f(A_1 \cup A_2) = f(A_1) \cup f(A_2), \tag{1.2.4}$$

but

$$f(A_1 \cap A_2) \subseteq f(A_1) \cap f(A_2). \tag{1.2.5}$$

The results (1.2.4), (1.2.5) extend to the union and intersection of any number of subsets of X; thus if $\{A_\lambda : \lambda \in \Lambda\}$ is a collection of subsets of X, then

$$f\Big(\bigcup_{\lambda\in\Lambda} A_\lambda\Big) = \bigcup_{\lambda\in\Lambda} f(A_\lambda), \tag{1.2.6}$$

and

$$f\Big(\bigcap_{\lambda\in\Lambda} A_\lambda\Big) \subseteq \bigcap_{\lambda\in\Lambda} f(A_\lambda). \tag{1.2.7}$$

The corresponding results concerning f^{-1} are better behaved.

If $A' \subseteq B' \subseteq X'$ then

$$f^{-1}(A') \subseteq f^{-1}(B'). \tag{1.2.8}$$

If $A_1', A_2' \subseteq X'$ then

$$f^{-1}(A_1' \cup A_2') = f^{-1}(A_1') \cup f^{-1}(A_2') \tag{1.2.9}$$

and

$$f^{-1}(A_1' \cap A_2') = f^{-1}(A_1') \cap f^{-1}(A_2'). \tag{1.2.10}$$

More generally if $\{A_\lambda' : \lambda \in \Lambda\}$ is a collection of subsets of X', then

$$f^{-1}\Big(\bigcup_{\lambda\in\Lambda} A_\lambda'\Big) = \bigcup_{\lambda\in\Lambda} f^{-1}(A_\lambda'), \tag{1.2.11}$$

and

$$f^{-1}\Big(\bigcap_{\lambda\in\Lambda} A_\lambda'\Big) = \bigcap_{\lambda\in\Lambda} f^{-1}(A_\lambda'). \tag{1.2.12}$$

For any set $A \subseteq X$,

$$f^{-1}\{f(A)\} \supseteq A; \tag{1.2.13}$$

there is equality for every set $A \subseteq X$ if and only if f is injective. For any set $A' \subseteq X'$,

$$f\{f^{-1}(A')\} \subseteq A'; \tag{1.2.14}$$

there is equality for every set $A' \subseteq X'$ if and only if f is surjective.

Finally, if $A' \subseteq X'$ then

$$f^{-1}(X'-A') = X-f^{-1}(A'). \tag{1.2.15}$$

1.3 Equivalence relations

Let X, X' be non-empty sets; then roughly speaking R is a relation from the elements of X to the elements of X' if, for each x in X and each x' in X', either x is related to x' in the manner prescribed by R, or it is not so related. More formally we may define a relation as follows.

Let X, X' be non-empty sets; if a rule is given, by which to each x in X there are assigned elements x' in X', then this rule is said to be a *relation from X to X'*. (Note that the rule may assign no elements of X' to a given x in X.) If the relation is denoted by R then we write xRx' if x is related to x'.

As for the definition of a function we can give the following alternative definition of a relation.

Let X, X' be non-empty sets; any subset of $X \times X'$ is said to be a relation from X to X'. If the subset is denoted by R, then we say that R is a relation from X to X'; we write xRx' if $(x, x') \in R$.

For our purposes either definition of a relation is acceptable and the reader should adopt whichever he prefers.

Henceforth we shall consider only relations R from a set X to itself; we then say that R is a *relation in X*.

Let R be a relation in X. If xRx for all x in X, R is said to be *reflexive*; if xRy implies yRx, R is said to be *symmetric*; if xRy and yRz imply xRz, R is said to be *transitive*. If R is a relation in X which is reflexive, symmetric and transitive, it is called an *equivalence relation*. In this case the symbol R is usually replaced by $\sim$.

Next we need the idea of a partition; this is straightforward. Let X be a non-empty set; then a class $\{X_\lambda : \lambda \in \Lambda\}$ of non-empty subsets is called a *partition* of X if

$$\bigcup_{\lambda \in \Lambda} X_\lambda = X$$

and, for any pair of elements λ, λ' of Λ, either $X_\lambda = X_{\lambda'}$ or $X_\lambda \cap X_{\lambda'} = \varnothing$. Thus each x in X belongs to one and only one of the distinct members of $\{X_\lambda : \lambda \in \Lambda\}$.

Let X be a non-empty set, and let $\sim$ be an equivalence relation defined in X; let $x \in X$. Then the set of all elements y in X such that $x \sim y$ is called the *equivalence class* of x with respect to $\sim$, and will be denoted by E_x; thus $E_x = \{y : x \sim y\}$.

We can now state the following important result.

THEOREM 1.3.1. *Let an equivalence relation* $\sim$ *be defined in the set* X; *then the collection* $\{E_x : x \in X\}$ *of all equivalence classes* (*with respect to* $\sim$) *is a partition of* X. *Furthermore, if* $x, y \in X$, *then* $E_x = E_y$ *if and only if* $x \sim y$.

For a discussion of equivalence relations and a proof of the above theorem see Herstein (1964), pp. 6–8.

1.4 Countability

A set X is said to be *countably infinite* if there exists a bijection from X onto N; if a set is either finite or countably infinite it is said to be *countable*. A set which is not countable (so must be infinite) is said to be *uncountable*.

Thus a set A is countable if and only if it can be written as a sequence $(a_1, a_2, \ldots)$ where each $a_i \in A$, and each element of A appears as an a_i.

The following results can be established.

(i) If A is a countably infinite set and B is an infinite subset of A, then B is also countably infinite.

(ii) The set of all rational numbers is countably infinite.

(iii) The set of all real numbers in $[0, 1]$ is uncountable.

(iv) The union of a countable collection of countable sets is also a countable set.

(v) If the sets $X_1, \ldots, X_n$ are countable, then so also is the set $X_1 \times \ldots \times X_n$.

1.5 The real number system

For our purposes the real number system is defined as any set of elements which satisfies the algebraic field axioms, the linear ordering axioms, and lastly the least upper bound axiom. These are listed in detail below. Starting from the natural numbers it can be shown, although it is a lengthy process, that there does exist a set R of elements which satisfies these axioms, and moreover R is essentially unique.

I. *The field axioms.*

A (i) For any x, y in R there is defined an element of R, called the *sum* of x, y and which will be denoted by $x+y$.

(ii) $x+y = y+x$ for all x, y in R.

(iii) $x+(y+z) = (x+y)+z$ for all x, y, z in R.

(iv) There exists a (unique) element of R, denoted by 0, such that $x+0 = x$ for all x in R.

(v) To each x in R there corresponds a (unique) element of R, denoted by $-x$, such that $x+(-x) = 0$.

B (i) For any x, y in R there is defined an element of R, called the *product* of x, y and which will be denoted by $x \,.\, y$ or xy.

(ii) $xy = yx$ for all x, y in R.

(iii) $x(yz) = (xy)z$ for all x, y, z in R.

(iv) There exists a (unique) element (not 0) of R, denoted by 1, such that $1 \,.\, x = x$ for all x in R.

(v) To each x in R such that $x \neq 0$, there corresponds a (unique) element of R, denoted by x^{-1} (or $1/x$), such that $x \,.\, x^{-1} = 1$.

C $\quad x(y+z) = xy+xz$ for all x, y, z in R.

II. *The linear ordering axioms.*

(i) There is defined a relation in R, denoted by $<$, such that for any x, y in R, one and only one of the possibilities $x < y$, $x = y$, $y < x$ holds.

(ii) If $x < y$ and $y < z$ then $x < z$.

(iii) If $x < y$ then $x+z < y+z$ for all z in R.

(iv) If $x < y$ and $0 < z$ then $xz < yz$.

If $x < y$ or $x = y$ we write $x \leqslant y$; if it is not true that $x < y$ then we write $x \nless y$. In view of (i), $x \leqslant y$ is equivalent to $y \nless x$. We also write $y \geqslant x$ to mean the same as $x \leqslant y$.

Before stating the final axiom for R, we need to define some terms. Let S be a non-empty set of real numbers.

If there exists M such that $x \leqslant M$ for all x in S, then M is said to be an *upper bound* of S and S is said to be *bounded above*. If there exists m such that $x \geqslant m$ for all x in S then m is said to be a *lower bound* of S and S is said to be *bounded below*. If S is bounded both above and below it is said to be *bounded*.

Now suppose that S is bounded above. If there exists K such that K is an upper bound of S, and $K \leqslant K'$ for any upper bound K' of S, then K is called the *least upper bound*, or *supremum*, of S. We denote it by lub S or sup S.

Similarly suppose that S is bounded below. If there exists k such that k is a lower bound of S, and $k \geqslant k'$ for any lower bound k' of S, then k is called the *greatest lower bound*, or *infimum*, of S. We denote it by glb S or inf S.

A question which naturally arises at this stage is, if S is a set of real numbers which is bounded above, does S necessarily possess a least upper bound? We would like the answer to this question to be 'yes' in order that the set R should have many properties which we intuitively expect of it; however that this is the case cannot be deduced from the above axioms for a linearly ordered field. Instead we must take the result (or some equivalent statement) to be our final axiom. It is this axiom which distinguishes R from a number of other sets which satisfy the axioms I, II.

We now formally state the last axiom.

III. *The least upper bound axiom.*

Any non-empty set of real numbers which is bounded above possesses a least upper bound.

It can be deduced from III that any non-empty set of real numbers which is bounded below possesses a greatest lower bound (and conversely).

Next we state some results concerning suprema and infima.

LEMMA 1.5.1. *Let S be a non-empty set of real numbers which is bounded above. Then $K =$ sup S if and only if*

(i) *$x \leqslant K$ for all x in S, and*

(ii) *given any $\varepsilon > 0$ there exists x in S such that $x > K - \varepsilon$.*

There is a similar equivalent characterization of the infimum of a set which is bounded below.

LEMMA 1.5.2. (i) *Let A, B be any two non-empty sets both of which are bounded above. Then $A \cup B$ is bounded above, and*

$$\sup (A \cup B) = \max (\sup A, \sup B).$$

(ii) *Let A, B be any two non-empty sets both of which are bounded below. Then $A \cup B$ is bounded below, and*

$$\inf (A \cup B) = \min (\inf A, \inf B).$$

There are no analogous results concerning the intersection of two sets.

LEMMA 1.5.3. *If A is bounded above and $\emptyset \subset B \subset A$, then B is bounded above, and* $\sup B \leqslant \sup A$. *If A is bounded below and $\emptyset \subset B \subset A$, then B is bounded below, and* $\inf A \leqslant \inf B$.

LEMMA 1.5.4. *Let A, B be any two non-empty sets and let*

$$A+B = \{a+b : a \in A, b \in B\}.$$

If A, B are bounded above, then so also is $A+B$ and

$$\sup (A+B) = \sup A + \sup B.$$

If A, B are bounded below, then so also is $A+B$ and

$$\inf (A+B) = \inf A + \inf B.$$

Let $f : X \to \mathsf{R}$ be a function such that the set $\{f(x) : x \in X\}$ is bounded above. Then $\sup \{f(x) : x \in X\}$ will also be written as

$$\sup_{x \in X} f(x),$$

or $\sup_x f(x)$ when no confusion can arise.

LEMMA 1.5.5. *Let f, g be two real-valued functions defined on a common domain X. If the sets*

$$\{f(x) : x \in X\}, \quad \{g(x) : x \in X\} \tag{1.5.1}$$

are bounded above, then so also is

$$\{f(x)+g(x) : x \in X\} \tag{1.5.2}$$

and

$$\sup_x \{f(x)+g(x)\} \leqslant \sup_x f(x) + \sup_x g(x). \tag{1.5.3}$$

If the sets (1.5.1) *are bounded below, then so also is* (1.5.2) *and*

$$\inf_x \{f(x)+g(x)\} \geqslant \inf_x f(x) + \inf_x g(x). \tag{1.5.4}$$

Strict inequality can occur in (1.5.3), (1.5.4). For example let $X = [0, 1]$ and let f, g be defined by $f(x) = x$, $g(x) = -x$; the details are left to the reader.

Note particularly the difference between the Lemmas 1.5.4 and 1.5.5.

1.6 Concerning continuous functions and Riemann integration

Throughout this section I will denote the closed interval $[a, b]$.

I. *Continuity.*

Let $f: X(\subseteq \mathsf{R}) \to \mathsf{R}$ be a function. Let $x_0 \in X$; if, given any $\varepsilon > 0$ there exists $\delta > 0$ for which $|f(x)-f(x_0)| < \varepsilon$ for all x in X such that $|x-x_0| < \delta$, then f is said to be *continuous at* x_0. If f is continuous at each point of X then f is said to be *continuous over* X.

If, given any $\varepsilon > 0$ there exists $\delta > 0$ for which $|f(x)-f(x')| < \varepsilon$ for all x, x' in X such that $|x-x'| < \delta$, where δ is independent of x, x', then f is said to be *uniformly continuous over* X.

THEOREM 1.6.1. *Suppose that* $f: I \to \mathsf{R}$ *is continuous over* I. *Then*

(i) f *is uniformly continuous over* I;

(ii) *the set* $f(I)$ *is bounded*;

(iii) *if* $M = \sup f(I)$, $m = \inf f(I)$, *and* $y \in [m, M]$, *there exists* x *in* I *such that* $f(x) = y$.

Let f, g be real-valued functions defined over a common domain X. Then we define the functions $|f|$, $\max(f, g)$, $\min(f, g)$ to be the real-valued functions defined over X given by

$$|f|(x) = |f(x)|$$

$$\max(f, g)(x) = \max(f(x), g(x)), \; \min(f, g)(x) = \min f((x), g(x)).$$

If X is a subset of R, and if f, g are continuous at some point of X, or over X, then so are also the functions $|f|$, $\max(f, g)$, $\min(f, g)$ continuous there. The continuity of $|f|$ follows from the inequality

$$\big||f(x)|-|f(y)|\big| \leqslant |f(x)-f(y)|.$$

The continuity of $\max(f, g)$ may then be deduced from the identity

$$\max(f(x), g(x)) = \tfrac{1}{2}\{|f(x)-g(x)|+(f(x)-g(x))\};$$

similarly for $\min(f, g)$.

If $f_1, \ldots, f_n$ are n real-valued functions defined on a common domain X, then the functions $\max(f_1, \ldots, f_n)$, $\min(f_1, \ldots, f_n)$ may be defined similarly.

It is convenient to include here the following definition.

Let X be an interval of the real line, and $f: X \to \mathsf{R}$ be a function. Let $x_0 \in X$; if, for some l in R, given any $\varepsilon > 0$ there exists $\delta > 0$ for which

$$\left| \frac{f(x)-f(x_0)}{x-x_0} - l \right| < \varepsilon$$

for all x in X such that $0 < |x-x_0| < \delta$, then f is said to be *differentiable at* x_0, and have *derivative* l there (usually denoted by $f'(x_0)$). If f is differentiable at each point of X then f is said to be *differentiable over* X.

II. *The definition of a Riemann integral.*

Let $f: I \to \mathsf{R}$ be a function which is bounded over I. Let P be a partition

$$a = x_0 < x_1 < x_2 < \ldots < x_n = b$$

of I; for $k = 1, \ldots, n$ let $\delta_k = x_k - x_{k-1}$, $I_k = [x_{k-1}, x_k]$,

$$M_k = \sup_{x \in I_k} f(x), \quad m_k = \inf_{x \in I_k} f(x),$$

and $$\bar{S}_P(f) = \textstyle\sum_P M_k \delta_k, \quad \underline{S}_P(f) = \sum_P m_k \delta_k,$$

where $\sum_P$ denotes summation from $k = 1$ to $k = n$. Then

$$\inf_P \bar{S}_P(f), \quad \sup_P \underline{S}_P(f)$$

both exist (since f is bounded), where the infimum and supremum are taken over the set of all partitions P of I; denote the infimum by j and the supremum by J. If j, J are equal, f is said to be *Riemann integrable* over I; the common value is called the *Riemann integral* of f over I and is denoted by

$$\int_a^b f(x)\,dx.$$

There are two very important classes of functions which are Riemann integrable.

THEOREM 1.6.2. *If $f: I \to \mathsf{R}$ is monotonic over I, or is continuous over I, then f is Riemann integrable over I.*

III. *The relation between integration and differentiation.*

THEOREM 1.6.3. *If* $f:I \to \mathsf{R}$ *is Riemann integrable over* I *then the function* $F:I \to \mathsf{R}$ *defined by*

$$F(x) = \int_a^x f(t)\,dt \qquad (a < x \leqslant b) \tag{1.6.1}$$

$(F(a) = 0)$ *is continuous over* I.

If, furthermore, f *is continuous over* I *then* F *is differentiable over* I *and* $F'(x) = f(x)$ *for all* x *in* I.

If $f:I \to \mathsf{R}$ is any Riemann integrable function, the function F defined by (1.6.1) will (also) be called the *definite integral* of f. If $g:I \to \mathsf{R}$ is a function and $G:I \to \mathsf{R}$ is a function differentiable over I such that $G'(x) = g(x)$ for all x in I, then G is called an *indefinite integral* of g.

The definite integral of a function over a given interval is unique (by its very manner of definition) but the indefinite integral is not unique (in both cases provided the integrals exist); however we do have the following elementary result. If $G:I \to \mathsf{R}$ is an indefinite integral of a function $g:I \to \mathsf{R}$, then any function $H:I \to \mathsf{R}$ of the form $H(x) = G(x)+C$, C a constant, is also an indefinite integral of g; furthermore every indefinite integral of g is of this form.

It is an important feature of Riemann integration that it is not exactly the reverse process of differentiation. Thus if f is Riemann integrable over I then it does not follow that the function F defined by (1.6.1) is differentiable; for example if f is given by

$$f(x) = 0 \text{ on } [a, c], \quad f(x) = 1 \text{ on } (c, b], \tag{1.6.2}$$

where $a < c < b$, then f possesses a definite integral F given by

$$F(x) = 0 \text{ on } [a, c], \quad F(x) = x-c \text{ on } (c, b], \tag{1.6.3}$$

but this function is not differentiable at c. In the reverse direction there exist functions which are differentiable but such that the derivative is not Riemann integrable.

However we have the following results.

THEOREM 1.6.4. *Let* $f:I \to \mathsf{R}$ *be differentiable over* I *and let* f' *be Riemann integrable. Then*

$$\int_a^b f'(x)\,dx = f(b)-f(a).$$

This may be reformulated as follows.

THEOREM 1.6.4′ (The fundamental theorem of calculus). *Let $f:I \to \mathrm{R}$ be Riemann integrable over I and let f possess an indefinite integral $G:I \to \mathrm{R}$. Then*

$$\int_a^b f(x)\,dx = G(b)-G(a).$$

In view of this result the function f defined by (1.6.2) cannot possess an indefinite integral, for otherwise the latter must differ by a constant from the function F defined by (1.6.3).

As a particularly important special case of Theorem 1.6.4′ it should be remembered that any function f which is continuous over I is both Riemann integrable there and also possesses an indefinite integral there.

IV. *Convergence and uniform convergence.*

If (x_n) is a real sequence and there exists x_0 for which, given any $\varepsilon > 0$, there exists N such that $|x_n - x_0| < \varepsilon$ for all $n > N$, then (x_n) is said to *converge to the limit x_0 as $n \to \infty$*; we write $x_n \to x_0$ as $n \to \infty$.

An important result is as follows.

THEOREM 1.6.5 (The Cauchy principle of convergence). *The real sequence (x_n) is convergent to some limit in R if and only if given any $\varepsilon > 0$ there exists N such that $|x_m - x_n| < \varepsilon$ for all $m, n > N$.*

THEOREM 1.6.6. *Let $f: X(\subseteq \mathrm{R}) \to \mathrm{R}$ and $x_0 \in X$. Then f is continuous at x_0 if and only if for any sequence (x_n) in X such that $x_n \to x_0$ then $f(x_n) \to f(x_0)$.*

Let (a_n) be a real sequence, and let $s_n = a_1 + \ldots + a_n$; if the sequence (s_n) is convergent to the limit s, then the infinite series

$$\sum_{n=1}^{\infty} a_n = a_1 + a_2 + \ldots$$

is said to *converge to s.*

When there can be no confusion, and for typographical reasons, we shall often write $\sum_n a_n$ in place of

$$\sum_{n=1}^{\infty} a_n.$$

Clearly Theorem 1.6.5 can be reformulated to give a result for the convergence of an infinite series.

If (x_n) is a sequence of functions $x_n: T(\subseteq \mathsf{R}) \to \mathsf{R}$, $n \in \mathsf{N}$, and there exists a function $x_0: T \to \mathsf{R}$ for which, given any $\varepsilon > 0$, there exists N such that $|x_n(t) - x_0(t)| < \varepsilon$ for all $n > N$ and all t in T where N is independent of t in T, then (x_n) is said to *converge to* x_0 *uniformly over* T.

Often we shall be interested in the case where T is a closed bounded interval. However it is stressed that the definition holds for any set T. Other cases which occur frequently are $T = \mathsf{N}$ and $T = \mathsf{R}$.

There are four basic results which we shall need concerning uniform convergence.

THEOREM 1.6.7. *Let* (x_n) *be a sequence of functions* $x_n: T(\subseteq \mathsf{R}) \to \mathsf{R}$, $n \in \mathsf{N}$, *and let* $x_0: T \to \mathsf{R}$. *Suppose that* $x_n \to x_0$ *uniformly over* T, *and* x_n *is continuous over* T *for each* n. *Then* x_0 *is continuous over* T.

THEOREM 1.6.8. *Let* (x_n) *be a sequence of functions* $x_n: I \to \mathsf{R}$, $n \in \mathsf{N}$, *and let* $x_0: I \to \mathsf{R}$. *Suppose that* $x_n \to x_0$ *uniformly over* I, *and* x_n *is Riemann integrable over* I *for each* n. *Then* x_0 *is Riemann integrable over* I *and moreover*

$$\lim_{n\to\infty} \int_a^b x_n(t)\,dt = \int_a^b x_0(t)\,dt.$$

Theorem 1.6.8 can be reformulated to give a result concerning the term-by-term integration of a uniformly convergent series.

THEOREM 1.6.9 (The Cauchy principle of uniform convergence). *The sequence* (x_n) *of functions* $x_n: T(\subseteq \mathsf{R}) \to \mathsf{R}$ *is uniformly convergent over* T *if and only if given any* $\varepsilon > 0$ *there exists* N *such that* $|x_n(t) - x_m(t)| < \varepsilon$ *for all* $m, n > N$ *and all* t *in* T *where* N *is independent of* t *in* T.

THEOREM 1.6.10 (The Weierstrass M-test). *Let* (x_n) *be a sequence of functions* $x_n: T(\subseteq \mathsf{R}) \to \mathsf{R}$ *and* (M_n) *be a real sequence such that* $|x_n(t)| \leqslant M_n$ *for all* t *in* T *and such that* $\sum_n M_n$ *is convergent. Then the series* $\sum_n x_n$ *is uniformly convergent over* T.

Next we state and prove a special result which will be required later.

THEOREM 1.6.11. *Let* $I = [a, b]$, $J = [c, d]$. *If*

(i) (g_n) *is a sequence of functions* $g_n: I \to J$ *which converge uniformly over* I *to a function* $g: I \to J$, *and*

(ii) *the function* $h: I \times J \to \mathsf{R}$ *is uniformly continuous over* $I \times J$, *then the sequence* (f_n) *of functions* $f_n: I \to \mathsf{R}$ *defined by* $f_n(t) = h(t, g_n(t))$ *converges uniformly over* I *to the function* $f: I \to \mathsf{R}$ *defined by* $f(t) = h(t, g(t))$.

Proof. Given any $\varepsilon > 0$, there exists $\delta > 0$ for which

$$|h(t, s) - h(t, s')| < \varepsilon$$

for all s, s' in J such that $|s - s'| < \delta$, where δ is independent of s, s'. Furthermore, for this δ, there exists N such that $|g_n(t) - g(t)| < \delta$ for all $n > N$ and all t in I, where N is independent of t. Hence

$$|h(t, g_n(t)) - h(t, g(t))| < \varepsilon$$

for all $n > N$ and all t in T.

1.7 Negation

Many proofs in mathematics are proofs by contradiction; thus in order to prove that some mathematical statement Q is true, it is supposed that Q is false, and from this some other statement which is obviously untrue is deduced. In many cases Q is a proposition which involves the quantifiers 'given any' (which is the same as 'for any' or 'for all') and/or 'there exists' (which is the same as 'for some').

For example, Q might be of the form 'given any $\varepsilon > 0$, $P(\varepsilon)$ is true', where $P(\varepsilon)$ is a statement involving ε (this appears, for example, in the definitions of limits and continuity). The negation of this is 'there exists at least one $\varepsilon > 0$ (say ε_0) such that $P(\varepsilon_0)$ is false'. It is essential to note that we need only one $\varepsilon > 0$ for which $P(\varepsilon)$ is false, and we do *not* require that $P(\varepsilon)$ be false for all $\varepsilon > 0$.

Another general form whose negation will be required is 'there exists N such that $P(N)$ is true'; the negation of this is 'for all N, $P(N)$ is false'.

These two negations are often all that are required to set up the negation of quite complicated mathematical statements; they may, however, have to be applied more than once in a given case.

As an example of this we construct the negation of 'given any $\varepsilon > 0$ there exists N such that $|a_n - a| < \varepsilon$ whenever $n > N$'. The negation of this is 'there exists at least one $\varepsilon > 0$ (say ε_0) such that it is false that there exists N such that $|a_n - a| < \varepsilon_0$ whenever $n > N$'; that is 'there exists at least one $\varepsilon > 0$ (say ε_0) such that for all N it is false that $|a_n - a| < \varepsilon_0$ whenever $n > N$'. Finally this is equivalent to 'there exists at least one $\varepsilon > 0$ (say ε_0) such that for all N, $|a_n - a| \geqslant \varepsilon_0$ for at least one $n > N$'.

But clearly 'for all N, $|a_n - a| \geqslant \varepsilon_0$ for at least one $n > N$' is equivalent to '$|a_n - a| \geqslant \varepsilon_0$ for an infinite number of values of n'. Thus we have established:

LEMMA 1.7.1. *The sequence (a_n) does not converge to a if (and only if) there exists at least one $\varepsilon > 0$ (say ε_0) such that, either*

(i) *for any N, there exists at least one $n > N$ for which $|a_n - a| \geqslant \varepsilon_0$,*

or

(ii) *$|a_n - a| \geqslant \varepsilon_0$ for an infinite number of values of n.*

Similar formulations can be made for the negations of statements involving uniform convergence and of continuity; for the latter we have

LEMMA 1.7.2. *The function $f: X(\subseteq \mathsf{R}) \to \mathsf{R}$ is not continuous at the point x_0 if and only if there exists at least one $\varepsilon > 0$ (say ε_0) such that, either*

(i) *for any $\delta > 0$, there exists at least one x for which $|x - x_0| < \delta$ and $|f(x) - f(x_0)| \geqslant \varepsilon_0$, or*

(ii) *$|f(x) - f(x_0)| \geqslant \varepsilon_0$ for some x arbitrarily close to x_0.*

1.8 Minkowski's inequalities

In this section we establish some inequalities due to Minkowski; in order to do this it is first necessary to derive some preliminary inequalities. Minkowski's inequalities will be required in Chapter 2 in some of the examples of metric spaces given there.

LEMMA 1.8.1. *If $p > 1$, q is defined by $1/p + 1/q = 1$ and $a \geqslant 0$, $b \geqslant 0$ then*

$$a^{1/p} b^{1/q} \leqslant \frac{a}{p} + \frac{b}{q}. \tag{1.8.1}$$

Proof. Let $f: [0, \infty) \to \mathsf{R}$ be defined by

$$f(t) = t^{\alpha} - \alpha t + \alpha - 1$$

where $0 < \alpha < 1$. Then $f'(t) = \alpha(t^{\alpha-1} - 1)$ so that $f(1) = 0, f'(1) = 0$,

$f'(t) > 0$ if $0 < t < 1$, and $f'(t) < 0$ if $t > 1$; hence $f(t) \leqslant 0$ for $t \geqslant 0$.

The inequality (1.8.1) clearly holds for $b = 0$, so assume $b > 0$. Set $t = a/b$ and $\alpha = 1/p$; then

$$f\left(\frac{a}{b}\right) = \left(\frac{a}{b}\right)^{1/p} - \frac{1}{p}\cdot\frac{a}{b} + \frac{1}{p} - 1 \leqslant 0.$$

Multiply by b and use $1/p+1/q = 1$; then (1.8.1) follows.

THEOREM 1.8.1 (Hölder's inequality for finite sums). *If* $p > 1$, q *is defined by* $1/p+1/q = 1$ *and* $x_1, \ldots, x_n$; $y_1, \ldots, y_n$ *are real (or complex) numbers, then*

$$\sum_{i=1}^{n} |x_i y_i| \leqslant \left\{\sum_{i=1}^{n} |x_i|^p\right\}^{1/p} \left\{\sum_{i=1}^{n} |y_i|^q\right\}^{1/q}.$$

Proof. For brevity set

$$X = \{\textstyle\sum_i |x_i|^p\}^{1/p}, \quad Y = \{\textstyle\sum_i |y_i|^q\}^{1/q}, \quad \text{where } \textstyle\sum_i = \displaystyle\sum_{i=1}^{n}.$$

Since the result is trivial if $X = 0$ or $Y = 0$ we assume that $X > 0$, $Y > 0$ and set

$$a_j = |x_j|^p/X^p, \quad b_j = |y_j|^q/Y^q, \quad j = 1, \ldots, n.$$

Then by Lemma 1.8.1 we have

$$\frac{|x_j y_j|}{XY} \leqslant \frac{a_j}{p} + \frac{b_j}{q};$$

hence

$$\frac{\sum_j |x_j y_j|}{XY} \leqslant \frac{1}{p}\{\textstyle\sum_j a_j\} + \frac{1}{q}\{\textstyle\sum_j b_j\} = \frac{1}{p} + \frac{1}{q} = 1.$$

When $p = q = 2$, Hölder's inequality reduces to

$$\left\{\sum_{i=1}^{n} |x_i y_i|\right\}^2 \leqslant \left\{\sum_{i=1}^{n} |x_i|^2\right\} \left\{\sum_{i=1}^{n} |y_i|^2\right\},$$

which is known as Cauchy's inequality.

We are now in a position to derive the two inequalities which we shall require.

THEOREM 1.8.2 (Minkowski's inequality for finite sums). *If $p \geqslant 1$ and $x_1, \ldots, x_n; y_1, \ldots, y_n$ are real (or complex) then*

$$\left\{\sum_{i=1}^{n} |x_i+y_i|^p\right\}^{1/p} \leqslant \left\{\sum_{i=1}^{n} |x_i|^p\right\}^{1/p} + \left\{\sum_{i=1}^{n} |y_i|^p\right\}^{1/p}. \quad (1.8.2)$$

Proof. First observe that (1.8.2) is self-evident when $p = 1$, so assume that $p > 1$. Clearly

$$\{\sum_i |x_i+y_i|^p\}^{1/p} \leqslant \{\sum_i(|x_i|+|y_i|)^p\}^{1/p};$$

but

$$(|x_i|+|y_i|)^p = |x_i|(|x_i|+|y_i|)^{p-1}+|y_i|(|x_i|+|y_i|)^{p-1}$$

so that

$$\sum_i(|x_i|+|y_i|)^p = \sum_i\{|x_i|(|x_i|+|y_i|)^{p-1}\} \\ +\sum_i\{|y_i|(|x_i|+|y_i|)^{p-1}\}. \quad (1.8.3)$$

Apply Hölder's inequality to the first sum on the right side of (1.8.3); it follows that

$$\sum_i\{|x_i|(|x_i|+|y_i|)^{p-1}\} \leqslant \{\sum_i|x_i|^p\}^{1/p}[\sum_i\{(|x_i|+|y_i|)^{p-1}\}^q]^{1/q} \\ = \{\sum_i|x_i|^p\}^{1/p}\{\sum_i(|x_i|+|y_i|)^p\}^{1/q}.$$

There is a similar result for the second sum on the right side of (1.8.3). Hence

$$\sum_i(|x_i|+|y_i|)^p \leqslant [\{\sum_i|x_i|^p\}^{1/p}+\{\sum_i|y_i|^p\}^{1/p}]\{\sum_i(|x_i|+|y_i|)^p\}^{1/q}.$$

Assume that x_i, y_i are not all zero (for then (1.8.2) is trivial); then

$$\sum_i(|x_i|+|y_i|)^p \neq 0,$$

and (1.8.2) follows.

Lastly we extend the result to (convergent) infinite sums.

THEOREM 1.8.3 (Minkowski's inequality for infinite sums). *If $p \geqslant 1$ and (x_n), (y_n) are real (or complex) sequences such that*

$$\sum_{i=1}^{\infty} |x_i|^p, \quad \sum_{i=1}^{\infty} |y_i|^p$$

are convergent, then

$$\sum_{i=1}^{\infty} |x_i+y_i|^p \quad (1.8.4)$$

is convergent, and moreover

$$\left\{\sum_{i=1}^{\infty}|x_i+y_i|^p\right\}^{1/p} \leqslant \left\{\sum_{i=1}^{\infty}|x_i|^p\right\}^{1/p} + \left\{\sum_{i=1}^{\infty}|y_i|^p\right\}^{1/p}. \qquad (1.8.5)$$

Proof. For any positive integer n we have

$$\left\{\sum_{i=1}^{n}|x_i+y_i|^p\right\}^{1/p} \leqslant \left\{\sum_{i=1}^{n}|x_i|^p\right\}^{1/p} + \left\{\sum_{i=1}^{n}|y_i|^p\right\}^{1/p}$$
$$\leqslant \left\{\sum_{i=1}^{\infty}|x_i|^p\right\}^{1/p} + \left\{\sum_{i=1}^{\infty}|y_i|^p\right\}^{1/p};$$

from this follows the convergence of (1.8.4) and the inequality (1.8.5).

2: BASIC TERMINOLOGY

2.1 Definition of a metric space

When studying the analytic properties of the set R of all real numbers it is clear that an important role is played by the distance function, that is, the function whose value is $|x-y|$ for x, y in R; similarly the corresponding distance function in the real plane R^2 (or any higher dimensional real space R^m, $m > 2$) is fundamental to the analytic study of R^2 (or R^m). In particular the very important concept of convergence depends essentially on the concept of distance. The three basic properties, which the distance function for R satisfies, are

(i) $|x-y| \geqslant 0$ for all x, y in R, with equality if and only if $x = y$;

(ii) $|x-y| = |y-x|$ for all x, y in R;

(iii) $|x-y| \leqslant |x-z|+|z-y|$ for all x, y, z in R.

There are, of course, exactly corresponding properties for R^2 or R^m.

There are many other sets of elements for which distance functions can be defined, that is, functions which satisfy properties of the same form as (i)–(iii); a number of explicit examples are given in §2.2, one of which we mention now. Let $I = [a, b]$, and let $\mathscr{C}(I)$ denote the set of all real-valued functions which are defined and continuous over the closed interval I; let $x, y \in \mathscr{C}(I)$ and call

$$\sup_{t \in I} |x(t)-y(t)|$$

the distance between x and y. Then it is clear that

$$\sup_t |x(t)-y(t)| \geqslant 0,$$

with equality if and only if $x(t) = y(t)$ for all t in I, that is $x = y$; trivially

$$\sup_t |x(t)-y(t)| = \sup_t |y(t)-x(t)|$$

for all x, y in $\mathscr{C}(I)$. Lastly using (iii) it is easily established that

$$\sup_t |x(t)-y(t)| \leqslant \sup_t |x(t)-z(t)|+\sup_t |z(t)-y(t)|$$

for all x, y, z in $\mathscr{C}(I)$. Thus we have a real-valued function, defined for all pairs of elements of $\mathscr{C}(I)$, which satisfies relations analogous to (i)–(iii). Thus we expect that all the properties of R which follow from (i)–(iii) alone, can be translated into results about $\mathscr{C}(I)$. This is indeed the case. In this way our knowledge of R is of great use in the study of $\mathscr{C}(I)$. We have specifically mentioned this set because of its obvious importance; however there are many other examples of sets for which a distance function satisfying relations analogous to (i)–(iii) can be defined.

Consequently it is of interest to investigate the properties that an arbitrary set X will possess if it is endowed with a distance function, but for which set no further assumptions are made, at least initially. A distance function as informally defined above is usually referred to as a metric function, which term will be used throughout this text.

We thus arrive at the following basic definition.

DEFINITION 2.1.1. Let X be a non-empty set. If there exists a mapping $\rho: X \times X \to \mathrm{R}$ such that

(i) $\rho(x, y) \geqslant 0$ for all x, y in X, with equality if and only if $x = y$,

(ii) $\rho(x, y) = \rho(y, x)$ for all x, y in X,

(iii) $\rho(x, y) \leqslant \rho(x, z) + \rho(z, y)$ for all x, y, z in X,

then the pair (X, ρ) is called a *metric space*, and the function ρ is called the *metric* of (X, ρ).

The elements of the set X are also called the points of the metric space (X, ρ). The axioms (i)–(iii) are called the metric space axioms; in particular, (ii) is the axiom of symmetry, and (iii) is the triangle inequality axiom.

Note that it is possible to associate with a set X more than one metric; if ρ_1, ρ_2 are two metrics for X then (X, ρ_1), (X, ρ_2) are two distinct metric spaces. In §2.2 we give examples of such situations.

Suppose that (X, ρ) is a metric space and Y is a non-empty subset of X. Then the restriction ρ_Y of the function ρ to the set $Y \times Y$ will serve as a metric for Y, since trivially it must satisfy the metric axioms over Y; thus (Y, ρ_Y) is also a metric space. (In accordance with the notation concerning restrictions introduced in §1.2, strictly we should have written $\rho_{Y \times Y}$ and not ρ_Y.) When no confusion can arise we may sometimes write (Y, ρ) instead of (Y, ρ_Y).

Probably the first results concerning metric spaces were those contained in Fréchet's doctoral thesis (1906).

EXERCISES 2.1

1. Let X be a non-empty set and $\rho: X \times X \to \mathrm{R}$ be a mapping such that

(*a*) $\rho(x, y) = 0$ if and only if $x = y$;

(*b*) $\rho(x, z) \leqslant \rho(y, x) + \rho(y, z)$ for all x, y, z in X.

Deduce that ρ is a metric on X. [Thus the three axioms for a metric space given in §2.1 could be replaced equivalently by two axioms, namely (*a*) and (*b*); we preferred not to do this in order to retain the conventional form for the axioms.]

2. Let $X = \mathrm{R}$ and define $\rho: \mathrm{R} \times \mathrm{R} \to \mathrm{R}$ by

$$\rho(x, y) = \begin{cases} |x-y|+1, & \text{if exactly one of } x, y \text{ is strictly positive,} \\ |x-y|, & \text{otherwise;} \end{cases}$$

prove that (X, ρ) is a metric space.

3. Define $\rho: \mathrm{R} \times \mathrm{R} \to \mathrm{R}$ by $\rho(x, y) = |x-y|^2$. Show that ρ is not a metric on R.

4. (i) Verify that $\lambda/(1+\lambda)$ is an increasing function of λ for $\lambda \geqslant 0$.

(ii) If λ, μ are non-negative real numbers show that

$$\frac{\lambda+\mu}{1+\lambda+\mu} \leqslant \frac{\lambda}{1+\lambda} + \frac{\mu}{1+\mu}.$$

(iii) Let (X, ρ) be a metric space; define $\rho': X \times X \to \mathrm{R}$ by

$$\rho'(x, y) = \frac{\rho(x, y)}{1+\rho(x, y)}.$$

Show that ρ' is also a metric on X.

5. Let $X = \mathrm{R}^2$ and define $\rho: X \times X \to \mathrm{R}$ by

$$\rho(x, y) = \begin{cases} \{(x_1-y_1)^2+(x_2-y_2)^2\}^{\frac{1}{2}}, & \text{if } x, y \text{ are collinear with the origin,} \\ (x_1^2+x_2^2)^{\frac{1}{2}}+(y_1^2+y_2^2)^{\frac{1}{2}}, & \text{otherwise,} \end{cases}$$

where $x = (x_1, x_2)$, $y = (y_1, y_2)$. Show that (X, ρ) is a metric space.

6. Let $X = \mathsf{R}^2$ and define $\rho: X \times X \to \mathsf{R}$ by

$$\rho(x, y) = \begin{cases} |x_2 - y_2| & \text{if } x_1 = y_1 \\ |x_1 - y_1| + |x_2| + |y_2| & \text{if } x_1 \neq y_1, \end{cases}$$

where $x = (x_1, x_2)$, $y = (y_1, y_2)$. Show that (X, ρ) is a metric space.

7. Let (X, ρ) be a metric space. Is (X, ρ^2) a metric space? Is $(X, \rho^{\frac{1}{2}})$ a metric space?

2.2 Examples of metric spaces

In the examples given below much of the verification of the metric space axioms is left to the reader.

(i) Let $X = \mathsf{R}$ and $\rho(x, y) = |x - y|$.

(ii) Let $X = \mathsf{R}^m$ (the set of real m-tuples); let $x = (x_1, \ldots, x_m)$, $y = (y_1, \ldots, y_m)$ and define

$$\rho(x, y) = \left\{ \sum_{i=1}^{m} |x_i - y_i|^2 \right\}^{\frac{1}{2}}.$$

To verify the third axiom, apply Minkowski's inequality (with $p = 2$) to the numbers $x_i - z_i$, $i = 1, \ldots, m$ and $z_i - y_i$, $i = 1, \ldots, m$.

In (i) and (ii) the metric is the 'usual' distance function; this will henceforth be called the *Euclidean* metric and be denoted by d (irrespective of the dimension m).

We now give some examples in which the metrics are rather different; first there is an easy generalization of (ii).

(iii) Let $X = \mathsf{R}^m$, and define

$$\rho(x, y) = \left\{ \sum_{i=1}^{m} |x_i - y_i|^p \right\}^{1/p}, \tag{2.2.1}$$

where $p \geqslant 1$. To verify the triangle inequality axiom, again use Minkowski's inequality.

(iv) Let $X = \mathsf{R}^m$, and now define

$$\rho(x, y) = \max_{1 \leq i \leq m} |x_i - y_i|.$$

The previous two examples illustrate the fact that there can exist more than one metric associated with a given set X; in (iii), for each different value of $p(\geqslant 1)$ we have a different metric, and thus we have an infinite number of metric spaces all with the same underlying set R^m.

(v) Let X be any non-empty set, and define

$$\rho(x, y) = \begin{cases} 0 & \text{if} \quad x = y, \\ 1 & \text{if} \quad x \neq y; \end{cases}$$

ρ is called the *standard discrete* metric for X.

(vi) Let $X = \ell^p$, the set of all real infinite sequences (x_i) such that

$$\sum_{i=1}^{\infty} |x_i|^p < \infty,$$

where $p \geqslant 1$. Let $x = (x_i), y = (y_i)$ be two elements of ℓ^p, and define

$$\rho(x, y) = \left\{\sum_{i=1}^{\infty} |x_i - y_i|^p\right\}^{1/p}.$$

It is stressed that for ρ to define a mapping of $X \times X$ into R it is necessary that $\rho(x, y) \in \mathsf{R}$ for all x, y in X. Hence in the present example the infinite series by means of which ρ is defined must be convergent for all x, y in X; for otherwise $\rho(x, y) \notin \mathsf{R}$. Now by Minkowski's inequality for infinite series we have

$$\rho(x, y) \leqslant \{\textstyle\sum_i |x_i|^p\}^{1/p} + \{\textstyle\sum_i |y_i|^p\}^{1/p} < \infty,$$

and so $\rho(x, y) \in \mathsf{R}$ for all x, y in X.

To verify the third metric axiom again use Minkowski's inequality.

(vii) Let $X = m$, the set of all bounded sequences, that is all infinite sequences (x_i) such that

$$\sup_{i \in \mathsf{N}} |x_i| < \infty,$$

and define

$$\rho(x, y) = \sup_{i \in \mathsf{N}} |x_i - y_i|.$$

Then

$$\rho(x, y) \leqslant \sup_i (|x_i| + |y_i|) \leqslant \sup_i |x_i| + \sup_i |y_i| < \infty,$$

and so $\rho(x, y) \in \mathsf{R}$ for all x, y in m.

Next three examples will be given in which the elements of X are functions; in each example I will denote the closed interval $[a, b]$.

(viii) Let $X = \mathscr{C}(I)$; for $x, y \in \mathscr{C}(I)$ define

$$\rho(x, y) = \sup_{t \in I} |x(t) - y(t)|. \tag{2.2.2}$$

It is left to the reader to show that $\rho(x, y) \in \mathsf{R}$ for all x, y in $\mathscr{C}(I)$; the verification that the axioms are satisfied was carried out in §2.1.

(ix) Let $X = \mathscr{B}(I)$, the set of all functions $x : I \to \mathsf{R}$ which are bounded over I; let $x, y \in \mathscr{B}(I)$, and define $\rho(x, y)$ as in (2.2.2). Then $(\mathscr{B}(I), \rho)$ is a metric space.

Of course the metric space of (viii) is a subspace of that of (ix).

(x) Let $X = \mathscr{C}(I)$, but now define

$$\rho(x, y) = \int_a^b |x(t) - y(t)|\, dt, \tag{2.2.3}$$

for any x, y in $\mathscr{C}(I)$.

If $x, y \in \mathscr{C}(I)$ then $|x - y| \in \mathscr{C}(I)$, the integral defining $\rho(x, y)$ is finite, and $(\mathscr{C}(I), \rho)$ is again a metric space.

In the above examples we could have taken our basic set to be the set C of complex numbers instead of R. Thus, for example, the set C^m with ρ given by (2.2.1) where x, y are now elements of C^m, defines a metric space. Likewise the set of all complex-valued functions which are defined and continuous over I with ρ defined by (2.2.2) or (2.2.3) form a metric space.

Having made this comment we shall not refer to these complex cases again but assume that the reader can make the necessary variations for himself.

Exercises 2.2

1. Let X be the set of all functions $x : I \to \mathsf{R}$ which are Riemann integrable over I and let $\rho : I \times I \to \mathsf{R}$ be defined by (2.2.3). Is ρ a metric on X?

2. Let $(X_i, \rho_i), i = 1, \ldots, m$ be metric spaces, and $X = X_1 \times \ldots \times X_m$; let $x = (x_1, \ldots, x_m)$, $y = (y_1, \ldots, y_m)$ where $x_i, y_i \in X_i$ and define

$$\rho(x, y) = \max_{1 \leq i \leq m} \rho_i(x_i, y_i), \quad \rho'(x, y) = \sum_{i=1}^{m} \rho_i(x_i, y_i).$$

Show that (X, ρ), (X, ρ') are metric spaces.

Discuss this in relation to the metric spaces given in (i)–(iv) of §2.2.

Give further metric functions that may be associated with the set X.

3. Let X be the set of all real sequences; let $x = (x_i)$, $y = (y_i)$ and define

$$\rho(x, y) = \sum_{i=1}^{\infty} \frac{1}{2^i} \frac{|x_i - y_i|}{1 + |x_i - y_i|}.$$

With the aid of Exercise 2.1.4, show that (X, ρ) is a metric space.

4. Let X be the set of all real sequences; let $x = (x_i)$, $y = (y_i)$, and define

$$\rho(x, y) = \sum_{i=1}^{\infty} \frac{1}{i^2} \min(|x_i - y_i|, 1).$$

Show that (X, ρ) is a metric space.

2.3 Convergence

It has already been mentioned that convergence of sequences in Euclidean space is a concept which depends only on 'distance'; we can therefore expect to be able to define convergence in general metric spaces. This is done very simply as follows.

DEFINITION 2.3.1. A sequence (x_n) of elements of a metric space (X, ρ) is said to *converge in* (X, ρ) if there exists x in X such that $\rho(x_n, x) \to 0$ as $n \to \infty$. We then write $x_n \to x$ as $n \to \infty$ or

$$\lim_{n \to \infty} x_n = x,$$

and call x the limit of the sequence (x_n).

If no confusion can arise we may omit reference to $n \to \infty$.

We can immediately deduce two results.

LEMMA 2.3.1. *If $x_n \to x$ and $y_n \to y$ in the metric space (X, ρ), then $\rho(x_n, y_n) \to \rho(x, y)$.*

Proof. For this we need the inequality

$$|\rho(x, y) - \rho(z, u)| \leqslant \rho(x, z) + \rho(y, u) \tag{2.3.1}$$

where $x, y, z, u \in X$. To prove (2.3.1), we have, by the triangle inequality

$$\rho(x, y) \leqslant \rho(x, z)+\rho(z, y) \leqslant \rho(x, z)+\rho(z, u)+\rho(u, y)$$

so that $$\rho(x, y)-\rho(z, u) \leqslant \rho(x, z)+\rho(u, y).$$

Similarly $$\rho(z, u)-\rho(x, y) \leqslant \rho(z, x)+\rho(y, u),$$

so combining these two results, (2.3.1) follows. Therefore

$$|\rho(x_n, y_n)-\rho(x, y)| \leqslant \rho(x_n, x)+\rho(y_n, y).$$

Since $x_n \to x$, $y_n \to y$, it follows that $\rho(x_n, y_n) \to \rho(x, y)$.

(We shall need the inequality (2.3.1) again.)

LEMMA 2.3.2. *If the sequence (x_n) converges in the metric space (X, ρ), then its limit is unique.*

Proof. Suppose that $x_n \to x$ and $x_n \to x'$ for some elements x, x' of X. By the triangle inequality

$$0 \leqslant \rho(x, x') \leqslant \rho(x, x_n)+\rho(x_n, x');$$

letting $n \to \infty$, it follows that $\rho(x, x') = 0$ so $x = x'$.

The classical definition of uniform convergence for sequences of real (or complex) valued functions is also readily extended to general metric spaces.

DEFINITION 2.3.2. Let (X, ρ) be a metric space, and let (f_n) be a sequence of functions defined on some set Y and taking values in X. If, for each y in Y, the sequence $(f_n(y))$ converges in (X, ρ), then the sequence (f_n) is said to *converge in (X, ρ) pointwise over Y.*

If (f_n) converges pointwise in (X, ρ), there exists a function $f_0: Y \to X$ such that $f_n(y) \to f_0(y)$ for each y in Y. Let the sequence (F_n) of functions $F_n: Y \to \mathsf{R}$ be defined by

$$F_n(y) = \rho(f_n(y), f_0(y));$$

if $F_n(y) \to 0$ uniformly over Y, in the classical sense, then the sequence (f_n) is said to *converge in (X, ρ) uniformly over Y.*

If there can be no ambiguity we may omit reference to either one or both of (X, ρ), Y.

Suppose that the sequence (f_n) is defined as in Definition 2.3.2; then it is uniformly convergent over Y if (and only if) given any

$\varepsilon > 0$, there exists N such that $\rho(f_n(y), f_0(y)) < \varepsilon$ for all $n > N$ and all y in Y, where N is independent of y.

Again suppose that (f_n) is defined as above, and that it is not necessarily uniformly convergent over Y. Let Y' be a subset of Y and g_n be the restriction of f_n to Y'; if (g_n) converges in (X, ρ) uniformly over Y', we may also say that the sequence (f_n) converges in (X, ρ) uniformly over Y'.

EXERCISES 2.3

1. Let (x_n) be a convergent sequence in a metric space (X, ρ), and let i be a fixed positive integer; show that $(\rho(x_i, x_n))$ is a bounded real sequence.

2. Let (x_n) be a sequence in a metric space (X, ρ) such that the three subsequences (x_{2n}), (x_{2n+1}), (x_{3n}) are all convergent. Deduce that (x_n) is convergent.

2.4 Examples concerning convergence in metric spaces

It is illustrative to examine how the definition of convergence in a metric space applies to some of the examples of §2.2; the Roman numerals below correspond to those of §2.2.

Throughout this text, whenever any metric space (X, ρ) is being considered in which the elements of X are either finite or infinite sequences, a sequence of points of X will be denoted by $(x^{(n)})$ instead of the usual (x_n); this is done to avoid confusion between the index n and the subscripts of the members of the individual sequences. In particular this will apply to examples (ii)–(iv), (vi), (vii) of §2.2.

(i), (ii) $X = \mathsf{R}$, R^m, and ρ is the corresponding Euclidean metric; then convergence in these metric spaces is, by definition, the classical concept of convergence. Henceforth a sequence which converges in a Euclidean metric space will be said to converge *in the Euclidean sense*.

(iii) $X = \mathsf{R}^m$, and

$$\rho(x, y) = \{\textstyle\sum_i |x_i - y_i|^p\}^{1/p}.$$

Let $x^{(n)} = (x_1^{(n)}, \ldots, x_m^{(n)})$; if $x^{(n)} \to x$ in (X, ρ), then

$$\{\textstyle\sum_i |x_i^{(n)} - x_i|^p\}^{1/p} \to 0,$$

so that $x_i^{(n)} \to x_i$, $i = 1, \ldots, m$. Conversely, if $x_i^{(n)} \to x_i$ for each i it follows that $x^{(n)} \to x$. Thus $x^{(n)} \to x$ in this metric space if and only if $x^{(n)} \to x$ in the Euclidean sense.

(iv) $X = \mathrm{R}^m$, and $\rho(x, y) = \max_i |x_i - y_i|$. It can again be shown that $x^{(n)} \to x$ in this metric space if and only if $x^{(n)} \to x$ in the Euclidean sense.

(v) X is any non-empty set, and ρ the standard discrete metric. Let (x_n) be any sequence in (X, ρ); then

$$\rho(x_n, x) = \begin{cases} 1 & \text{if } x_n \neq x, \\ 0 & \text{if } x_n = x. \end{cases}$$

Thus $x_n \to x$ in (X, ρ) if and only if all the x_n, except possibly for an initial finite number, are equal to x.

(vi) $X = \ell^p$, and

$$\rho(x, y) = \{\textstyle\sum_i |x_i - y_i|^p\}^{1/p};$$

let $x^{(n)} = (x_1^{(n)}, x_2^{(n)}, \ldots)$ so that

$$\rho(x^{(n)}, x) = \{\textstyle\sum_i |x_i^{(n)} - x_i|^p\}^{1/p}.$$

If $x^{(n)} \to x$ in (X, ρ) then, given any $\varepsilon > 0$, there exists N such that

$$\textstyle\sum_i |x_i^{(n)} - x_i|^p < \varepsilon^p \qquad \text{for all } n > N,$$

so that $|x_i^{(n)} - x_i|^p < \varepsilon^p$ for all $n > N$, $i = 1, 2, \ldots$ and thus $x_i^{(n)} \to x_i$ $(i = 1, 2, \ldots)$ in the Euclidean sense.

However the converse is not true; for example, consider the sequence $x_i^{(n)} = x_i + \delta_{in}$, where

$$\delta_{in} = \begin{cases} 1 & \text{if } i = n, \\ 0 & \text{if } i \neq n. \end{cases}$$

(vii) $X = m$, and $\rho(x, y) = \sup_i |x_i - y_i|$; let $x^{(n)} = (x_1^{(n)}, x_2^{(n)}, \ldots)$ so that $\rho(x^{(n)}, x) = \sup_i |x_i^{(n)} - x_i|$. If $x^{(n)} \to x$ in this metric space then, given any $\varepsilon > 0$, there exists N such that

$$\sup_i |x_i^{(n)} - x_i| < \varepsilon \qquad \text{for all } n > N, \tag{2.4.1}$$

so

$$|x_i^{(n)} - x_i| < \varepsilon \qquad \text{for all } n > N, \quad i = 1, 2, \ldots, \tag{2.4.2}$$

and thus $x_i^{(n)} \to x_i$ $(i = 1, 2, \ldots)$ in the Euclidean sense. In fact this convergence is uniform with respect to i; for N is independent of i in (2.4.1), so the same is true in (2.4.2).

Conversely if the sequences $(x_i^{(n)})$ converge in the Euclidean sense as $n \to \infty$, uniformly in i, and $x_i^{(n)} \to x_i$, say, then $x^{(n)} \to x$ in (X, ρ) as $n \to \infty$. This follows immediately by reversing the above steps.

(viii) $X = \mathscr{C}(I)$, and $\rho(x, y) = \sup_t |x(t)-y(t)|$. Suppose $x_n \to x$ in (X, ρ); then given any $\varepsilon > 0$, there exists N such that

$$\sup_t |x_n(t)-x(t)| < \varepsilon \qquad \text{for all } n > N,$$

so $|x_n(t)-x(t)| < \varepsilon$ for all $n > N$, $t \in I$, where N is independent of t. Thus $x_n \to x$ uniformly over I in the classical sense which, again, will be referred to as the Euclidean sense.

Conversely if the sequence (x_n) converges, as $n \to \infty$, uniformly over I in the Euclidean sense, where $x_n \in \mathscr{C}(I)$, then (x_n) converges in $(\mathscr{C}(I), \rho)$. This follows immediately by reversing the above steps, and using Theorem 1.6.9.

(ix) $X = \mathscr{B}(I)$, and ρ is defined by the supremum metric as in (viii). Then, as in (viii), $x_n \to x$ in this metric space if and only if $x_n \to x$ uniformly over I, in the Euclidean sense.

Thus (x_n) is a convergent sequence in $(\mathscr{B}(I), \rho)$ if and only if (x_n) is convergent (in the Euclidean sense) uniformly over I. For this reason the supremum metric is sometimes called the *uniform* metric.

(x) $X = \mathscr{C}(I)$, and

$$\rho(x, y) = \int_a^b |x(t)-y(t)|\, dt.$$

Thus, if $x_n \to x$ in this metric space, then

$$\int_a^b |x_n(t)-x(t)|\, dt \to 0 \qquad \text{as } n \to \infty.$$

We give an example to show that convergence in this metric space does *not* imply pointwise convergence in the Euclidean sense.

Define a sequence (x_n) in $\mathscr{C}(I)$ by

$$x_n(t) = \begin{cases} n(t-a) & \text{for } a \leqslant t \leqslant a+1/n \\ 1 & \text{for } a+1/n < t \leqslant b; \end{cases}$$

let x be defined by $x(t) = 1$ for $a \leqslant t \leqslant b$. Then $\rho(x_n, x) = (2n)^{-1}$ so $\rho(x_n, x) \to 0$ as $n \to \infty$; that is $x_n \to x$ in this metric space. But $x_n(a) = 0$, so x_n does not converge to x pointwise in the Euclidean sense over I. This establishes our assertion.

On the other hand if (x_n) converges, in the Euclidean sense, to x uniformly over I, where each x_n is continuous over I, then $x_n \to x$ in the metric space (X, ρ). The verification of this assertion is left to the reader.

The above examples show that convergence in a metric space (X, ρ) may mean substantially different things in different spaces arising from the same set X.

EXERCISES 2.4

1. With the notation of Exercise 2.2.2, let $x^{(n)} = (x_1^{(n)}, \ldots, x_m^{(n)})$; show that $x^{(n)} \to x$ in (X, ρ), or in (X, ρ'), if and only if $x_i^{(n)} \to x_i$ in (X_i, ρ_i) for $i = 1, \ldots, m$.

2. With the notation of Exercise 2.2.3, let $x^{(n)} = (x_i^{(n)})$; show that $x^{(n)} \to x$ in (X, ρ) if and only if $x_i^{(n)} \to x_i$ for all i.
Establish the same result for the metric space defined in Exercise 2.2.4.

3. Describe the sequences which converge in the metric spaces defined in Exercises 2.1.5, 2.1.6.

4. Let $X = \mathscr{C}(\mathsf{R})$, the set of all real-valued functions which are continuous over R. Let $x, y \in X$; define

$$\rho_i(x, y) = \sup_{|t| \leq i} |x(t) - y(t)|,$$

and

$$\rho(x, y) = \sum_{i=1}^{\infty} \frac{1}{2^i} \cdot \frac{\rho_i(x, y)}{1 + \rho_i(x, y)}.$$

Show that (X, ρ) is a metric space, and that the sequence (x_n) converges to x in (X, ρ) if and only if $x_n \to x$ in the Euclidean sense uniformly over $[-i, i]$ for all $i > 0$.

2.5 The basic topological concepts

The concepts of neighbourhood, limit point, open set and closed set will be defined for metric spaces; these are usually called topological concepts since they are meaningful in (and moreover fundamental to) certain spaces, known as topological spaces, which are of a more general nature than metric spaces.

Definition 2.5.1. Let (X, ρ) be a metric space; given any x_0 in X and any $r > 0$, the set of all points x in X such that $\rho(x_0, x) < r$ is called an *open sphere* in (X, ρ), and is denoted by $S(x_0, r)$.

Correspondingly, the set of all points x in X such that $\rho(x_0, x) \leqslant r$ is called a *closed sphere* in (X, ρ) and is denoted by $\bar{S}(x_0, r)$.

In either case, the point x_0 is called the *centre* of the sphere and r the *radius* of the sphere.

Definition 2.5.2. Let (X, ρ) be a metric space; a *neighbourhood* of the point x_0 in X is any open sphere in (X, ρ) with centre x_0.

Some authors call this a spherical neighbourhood of x_0; they then define a general neighbourhood of x_0 as any set which contains a spherical neighbourhood of x_0. This sophistication will not be required in the present text.

Lemma 2.5.1. *Let (X, ρ) be a metric space and $x \in X$. If $S(x, \varepsilon)$ is any neighbourhood of x and $y \in S(x, \varepsilon)$, then there exists a neighbourhood $S(y, \varepsilon')$ of y such that $S(y, \varepsilon') \subseteq S(x, \varepsilon)$.*

Proof. Let $\varepsilon' = \varepsilon - \rho(x, y)$ (clearly $\rho(x, y) < \varepsilon$); then it is easily verified that $S(y, \varepsilon') \subseteq S(x, \varepsilon)$.

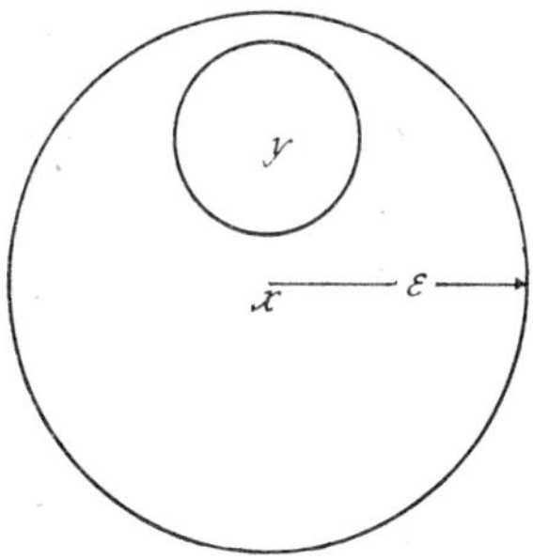

The diagram only suggests a method of proof; however it can do nothing more since it portrays the case $X = R^2$ with ρ being the Euclidean metric.

Definition 2.5.3. Let (X, ρ) be a metric space, and $Y \subseteq X$; a point x_0 in Y is called an *interior point* of Y if there exists a neighbourhood of x_0 lying entirely in Y.

It is seen that this implies that every point of X is an interior point of X, which fact we shall return to in the next section.

DEFINITION 2.5.4. Let (X, ρ) be a metric space, and $Y \subseteq X$; a point x_0 in X is called a *limit point* of Y if every neighbourhood of x_0 contains at least one point ($\neq x_0$) of Y.

Thus x_0 is a limit point of Y if $(S(x_0, \varepsilon)-\{x_0\}) \cap Y \neq \varnothing$ for all $\varepsilon > 0$. Some writers use the term 'point of accumulation' instead of limit point.

We give some simple examples and make some remarks.

(i) In $(\mathsf{R}^2, \mathsf{d})$ let $Y_1 = S(0, 1)$, $Y_2 = \bar{S}(0, 1)$, where $0 = (0, 0)$. Then every point of Y_1 is an interior point of Y_1 and is also an interior point of Y_2; every point of Y_2 is a limit point of Y_1 and is also a limit point of Y_2.

(ii) For any subset Y of the discrete space (X, ρ) defined in §2.2 (v), every point of Y is an interior point of Y, but no point of X is a limit point of Y; why?

(iii) It is stressed that a limit point of a set does not necessarily belong to that set; for example see the set Y_1 defined in (i).

(iv) A subset Y of a metric space (X, ρ) may contain points which are not limit points of Y; any such point is called an *isolated point* of Y. A point x_0 of Y is an isolated point of Y if and only if there exists $\varepsilon > 0$ such that

$$(S(x_0, \varepsilon)-\{x_0\}) \cap Y = \varnothing.$$

(v) The existence of one point as specified in Definition 2.5.4 implies the existence of an infinite number of such points. For, if ε is the radius of an arbitrary neighbourhood of x_0, which contains a point x_1 (say) of Y other than x_0, let $\varepsilon_1 = \rho(x_0, x_1) < \varepsilon$. Then the neighbourhood $S(x_0, \varepsilon_1)$ of x_0 also contains a point x_2 (say) of Y other than x_0, and since $x_1 \notin S(x_0, \varepsilon_1)$, it follows that $x_1 \neq x_2$; let $\varepsilon_2 = \rho(x_0, x_2)$. By next considering the neighbourhood $S(x_0, \varepsilon_2)$ we obtain a third point x_3 distinct from x_0, x_1, x_2; and so on.

Thus in Definition 2.5.4 'at least one point ($\neq x_0$) of Y' could be replaced equivalently by 'an infinite number of points of Y'.

(vi) It is trivial that no finite subset of a metric space possesses a limit point.

The next result gives an equivalent characterization of limit points.

THEOREM 2.5.1. *Let (X, ρ) be a metric space and $Y \subseteq X$; then the following statements are equivalent*:

(i) *x_0 is a limit point of Y*;

(ii) *there exists a sequence* (x_n) *of distinct points of* Y *such that* $x_n \to x_0$.

Proof. The proof that (ii) implies (i) is trivial.

Now assume (i); given any $\varepsilon_0 > 0$, there exists a point x_1 (say) of Y in $S(x_0, \varepsilon_0)$ where $x_1 \neq x_0$. Let $\varepsilon_1 = \frac{1}{2}\rho(x_0, x_1) < \frac{1}{2}\varepsilon_0$; then there exists a point x_2 (say) of Y in $S(x_0, \varepsilon_1)$ where $x_2 \neq x_0$. Clearly $x_2 \neq x_1$. Let $\varepsilon_2 = \frac{1}{2}\rho(x_0, x_2) < \frac{1}{2}\varepsilon_1$; proceeding in this way (that is, by induction) we obtain an infinite sequence of distinct points such that

$$\rho(x_0, x_n) < \varepsilon_{n-1} < \tfrac{1}{2}\varepsilon_{n-2} < \ldots < \frac{1}{2^{n-1}}\varepsilon_0.$$

Then $\rho(x_0, x_n) \to 0$ as $n \to \infty$, so $x_n \to x_0$.

The ideas of a limit and a limit point are distinct and apply in different situations; thus a limit is a point associated with a sequence, whereas a limit point is a point associated with a set. Thus a sequence may possess a limit but will never possess a limit point; on the other hand a set may possess (one or more) limit points but it will never possess a limit. We can, however, establish some connection between these ideas, when given a sequence, by considering the limit points (if any) of the set whose elements are the members of the sequence.

LEMMA 2.5.2. *Let* (X, ρ) *be a metric space containing a sequence which converges to a point* x; *then* x *is a limit point of the set whose elements are the points of the sequence if and only if the sequence has infinitely many distinct points.*

Proof. This is left as an exercise.

EXERCISES 2.5

1. Let $X = \{(x_1, x_2): x_1^2 + x_2^2 \leqslant 1, x_1, x_2 \in \mathsf{R}\}$; in the metric space (X, d), describe which points belong to the open spheres having (i) centre (0, 0), radii $\frac{1}{2}$, 1, (ii) centre (1, 0), radius $\frac{1}{4}$.

2. Let $X = \{(x_1, x_2): |x_1| < 1, |x_2| < 2, x_1, x_2 \in \mathsf{R}\}$; in the metric space (X, d) show that $S(x, r) = X$ for all $r \geqslant 2\sqrt{5}$ and all x in X.

3. Let $X = \{(x_1, x_2): x_1^2 + x_2^2 < 9, x_1, x_2 \in \mathsf{R}\}, x = (0, 0), x' = (2, 0)$. In the metric space (X, d) show that $S(x', 4) \subset S(x, 3)$. [Thus in an arbitrary metric space $S(x', r') \subset S(x, r)$ does not necessarily imply that $r' \leqslant r$.]

4. Give an example of a metric space (X, ρ) and a non-empty subset Y of X with the property that no point of Y is either an interior point or a limit point of Y.

5. Describe the form of a sphere in each of the metric spaces defined in §2.2 (iv), (v), (viii).

6. Describe the form of a sphere in each of the metric spaces defined in Exercises 2.1.5, 2.1.6.

2.6 Open and closed sets: introduction

The basic definitions of the section are as follows.

DEFINITION 2.6.1. Let (X, ρ) be a metric space, and $Y \subseteq X$. Then Y is said to be *open in* (X, ρ) if every point of Y is an interior point of Y.

DEFINITION 2.6.2. Let (X, ρ) be a metric space, and $Y \subseteq X$. Then Y is said to be *closed in* (X, ρ) if its complement $X - Y$ is open in (X, ρ).

If no confusion can arise, it will just be said that Y is open or Y is closed respectively.

The following points should be noted.

(i) Definition 2.5.1 is compatible with the above definitions; that is, an open sphere is an open set and a closed sphere is a closed set. The proof of these statements is left as an exercise.

(ii) A given set $Y \subseteq X$ is not necessarily open or closed: for example if $Y = (a, b]$ then Y is neither open nor closed in (R, d).

(iii) A given set $Y \subseteq X$ may be open and closed simultaneously. For example, the sets $\varnothing$ and X are both open and closed in any metric space (X, ρ). To establish this, note that a set Y is open if all its points are interior points of Y; since $\varnothing$ contains no points, it certainly contains no points which are *not* interior points, so $\varnothing$ is open in (X, ρ). By the comment immediately following Definition 2.5.3, it follows that X is open in (X, ρ). By the definition of a closed set, the two facts just proved imply, respectively, that X and $\varnothing$ are closed in (X, ρ).

In view of (ii), (iii) sets are said to be 'unlike doors', since a set may be simultaneously open and closed, or alternatively it may be neither open nor closed.

(iv) If Y is open in (X, ρ), then $X-Y$ is closed in (X, ρ); the proof of this is left to the reader as an (easy) exercise.

The following question arises from (iii): are there any other (so proper) subsets of (X, ρ) which are simultaneously open and closed? The full answer to this is 'sometimes yes, and sometimes no, depending on the nature of (X, ρ)'; this will be discussed in depth later (Chapter VI). However we will now give an example to show that the answer is sometimes yes.

Let $X = \{x : |x| \geqslant 1, x \in \mathsf{R}\}$; then it is easily verified that $[1,\infty)$ is a proper subset of X which is open and closed in (X, d).

We now come to an important result.

THEOREM 2.6.1. *Let (X, ρ) be a metric space and $Y \subseteq X$; then the following statements are equivalent*:

(i) *Y is closed*;

(ii) *Y contains all its limit points.*

Proof. Assume (i), so $X-Y$ is open. Let $x \in X-Y$ so there exists a neighbourhood $S = S(x, \varepsilon)$ of x such that $S \subseteq X-Y$. Thus $S \cap Y = \varnothing$ and hence $(S-\{x\}) \cap Y = \varnothing$. Therefore x is not a limit point of Y, so (i) implies (ii).

Now assume (ii); we show that $X-Y$ is open. Let $x \in X-Y$, so $x \notin Y$. Thus x is not a limit point of Y, so there exists a neighbourhood $S = S(x, \varepsilon)$ of x such that $(S-\{x\}) \cap Y = \varnothing$; since $x \notin Y$, it follows that $S \cap Y = \varnothing$, and so $S \subseteq X-Y$. Thus x is an interior point of $X-Y$; hence $X-Y$ is open.

DEFINITION 2.6.3. Let (X, ρ) be a metric space and $Y \subseteq X$; the set consisting of Y together with all its limit points is called the *closure* of Y, and is denoted by $\bar{Y}$.

It is obvious, but none-the-less important to notice, that

(i) $Y \subseteq \bar{Y}$ for all sets $Y \subseteq X$;

(ii) $Y = \bar{Y}$ if and only if Y is closed.

LEMMA 2.6.1. *Let (X, ρ) be a metric space and $Y \subseteq X$. Then the following statements are equivalent*:

(i) $x \in \bar{Y}$;

(ii) *$S(x, \varepsilon) \cap Y \neq \varnothing$ for every neighbourhood $S(x, \varepsilon)$ of x*;

(iii) *there exists an infinite sequence* (x_n) *of points (not necessarily distinct) of* Y *such that* $x_n \to x$.

Proof. This is left as an exercise.

THEOREM 2.6.2. *Let* (X, ρ) *be any metric space, and* $Y \subseteq X$; *let* Y' *denote the set of all limit points of* Y. *Then*

(i) $\bar{Y}$ *is closed (so* $\bar{\bar{Y}} = \bar{Y}$);

(ii) Y' *is closed.*

Proof. (i) Let x be a limit point of $\bar{Y}$ and let $S = S(x, \varepsilon)$ be any neighbourhood of x; then S contains a point y of $\bar{Y}$. By Lemma 2.5.1, S contains some neighbourhood of y, and therefore some point of Y (by Lemma 2.6.1). Thus $S \cap Y \neq \varnothing$ for every neighbourhood of x, and so $x \in \bar{Y}$. Hence $\bar{Y}$ is closed.

(ii) Now let x be a limit point of Y' and proceed as before. Thus any neighbourhood S of x contains a point y of Y' and hence an infinity of points of Y (see the remark (v) following Definition 2.5.4). Therefore $x \in Y'$ and so Y' is closed.

In view of the remark (ii) of §2.5, if X is any non-empty set and ρ is the standard discrete metric on X, then every subset of (X, ρ) is open; hence every subset of (X, ρ) is closed. This motivates

DEFINITION 2.6.4. Any metric space in which every subset is open is said to be *discrete*; the metric of a discrete space is called a *discrete metric*.

The proof of the next result is left as another exercise.

LEMMA 2.6.2. *Let* (X, ρ) *be a metric space. The following statements are equivalent*:

(i) (X, ρ) *is discrete*;

(ii) *every subset of* (X, ρ) *is closed*;

(iii) *for each* x *in* X *there exists* $\varepsilon > 0$ *such that* $S(x, \varepsilon) = \{x\}$, *so* $\{x\}$ *is open*;

(iv) *every point of* (X, ρ) *is an isolated point*;

(v) *the only convergent sequences in* (X, ρ) *are those which are eventually constant*;

(vi) *every subset of* (X, ρ) *has no limit point.*

EXERCISES 2.6

1. Let (X, d) be the metric space defined in Exercise 2.5.1, and let Y_1, Y_2, Y_3 be the subsets of X defined by

$$Y_1 = \{(x_1, x_2) : \sqrt{(x_1^2+x_2^2)} < \tfrac{1}{2}\},$$

$$Y_2 = \{(x_1, x_2) : \tfrac{1}{2} < \sqrt{(x_1^2+x_2^2)} \leqslant 1\},$$

$$Y_3 = \{(x_1, x_2) : \tfrac{1}{2} < \sqrt{(x_1^2+x_2^2)} < 1\}.$$

Which of Y_1, Y_2, Y_3 are open or closed in (X, d)?

2. Let

$$A = \{(x_1, x_2) : (x_1+1)^2+x_2^2 \leqslant 1, x_1, x_2 \in \mathrm{R}\}$$

$$B = \{(x_1, x_2) : (x_1-1)^2+x_2^2 < 1, x_1, x_2, \in \mathrm{R}\};$$

what is $\bar{B}$ in the space $(A \cup B, \mathrm{d})$?

3. Let $A \subseteq I\,(= [a, b])$; let $\mathscr{A}$ denote the subset of $\mathscr{C}(I)$ defined by

$$\mathscr{A} = \{f : f(t) = 0 \text{ for all } t \text{ in } A\}.$$

Show that $\mathscr{A}$ is a closed subset of $(\mathscr{C}(I), \rho)$ when ρ is the supremum metric (of §2.2(viii)), but $\mathscr{A}$ is not necessarily closed when ρ is the integral metric (of §2.2(x)). [To establish the second assertion consider the example given in §2.4(x).]

4. Let (X, ρ) be any metric space; prove that $\overline{S(x, r)} \subseteq \bar{S}(x, r)$ for all x in X and all $r > 0$. Give an example to show that strict inclusion is possible.

5. If Y is an open subset of (X, ρ) and $x_0 \in Y$ then prove that $Y-\{x_0\}$ is also open. Deduce that if any finite number of points are removed from Y then the remaining set is open.

Does this assertion extend to the removal of an infinite number of points?

6. Let (m, ρ) be the metric space defined in §2.2 (vii), and let $c \subset m$ be the subset consisting of all convergent sequences. Show that c is closed in (m, ρ).

2.7 Open and closed sets: continued

The basic results concerning open sets are contained in the next theorem.

THEOREM 2.7.1. *Let* (X, ρ) *be a metric space. Then*

(i) X, $\varnothing$ *are open in* (X, ρ);

(ii) *the union of any collection of open sets is open*;

(iii) *the intersection of any finite collection of open sets is open.*

Proof. (i) This was proved in §2.6.

(ii) Let $\{Y_\lambda : \lambda \in \Lambda\}$ be a collection of open sets in (X, ρ), and denote the union of all the sets Y_λ by Y. Let $x \in Y$; then there exists some λ in Λ (say λ_0) such that $x \in Y_{\lambda_0}$. Since Y_{λ_0} is open in (X, ρ), there exists some neighbourhood $S = S(x, \varepsilon)$ of x such that $S \subseteq Y_{\lambda_0}$; hence $S \subseteq Y$. Thus Y is open in (X, ρ).

(iii) First it will be shown that the intersection of two open sets Y_1, Y_2 is open. Assume that $Y_1 \cap Y_2 \neq \varnothing$ (for if $Y_1 \cap Y_2 = \varnothing$ there is nothing to prove); let $x \in Y_1 \cap Y_2$, so $x \in Y_1$ and $x \in Y_2$. Then there exist neighbourhoods $S(x, \varepsilon_1)$, $S(x, \varepsilon_2)$ of x such that $S(x, \varepsilon_1) \subseteq Y_1$ and $S(x, \varepsilon_2) \subseteq Y_2$. Let $\varepsilon = \min(\varepsilon_1, \varepsilon_2)$; then $S(x, \varepsilon) \subseteq Y_1 \cap Y_2$ and hence $Y_1 \cap Y_2$ is open.

By induction it follows that the intersection of any finite collection of open sets is open.

To see that the intersection of an infinite collection of open sets is not necessarily open, consider the following example. Let $I_n = (0, 1+n^{-1})$ for $n \in \mathsf{N}$, so that I_n is open in (R, d); then

$$\bigcap_{n=1}^{\infty} I_n = (0, 1],$$

which is not open in (R, d).

There is a corresponding set of results for closed sets.

THEOREM 2.7.2. *Let* (X, ρ) *be a metric space. Then*

(i) X, $\varnothing$ *are closed in* (X, ρ);

(ii) *the intersection of any collection of closed sets is closed*;

(iii) *the union of any finite collection of closed sets is closed.*

Proof. Again (i) was proved in §2.6.

Since $Y \subseteq X$ is closed if and only if $X - Y$ is open, (ii) and (iii) may be deduced from (ii) and (iii) of Theorem 2.7.1 by using De Morgan's rules. The details are left as an exercise.

Notice particularly that although the union of any collection of open sets is open, the corresponding result concerning the union of closed sets requires that the collection be finite; the reverse statement holds for intersections. The above counterexample may be used as an aid to remembering which way round these results hold.

It has been mentioned previously that by noticing the important role played by the function which is the distance between elements in R we are led to the study of sets of elements possessing only a distance function (or metric). Likewise, in many problems concerning metric spaces, it can be seen that only a minor role is played by the metric itself, and that the concept of openness (or nearness) plays a more important part. This leads to a generalization of metric spaces, namely to the study of sets satisfying another group of axioms—those which are just the results (i)–(iii) of Theorem 2.7.1. Thus we have

DEFINITION 2.7.1. Let X be a non-empty set. If there exists a collection $\mathscr{T}$ of subsets of X such that

(i) $X, \varnothing \in \mathscr{T}$,
(ii) the union of any collection of sets in $\mathscr{T}$ is in $\mathscr{T}$,
(iii) the intersection of any finite collection of sets in $\mathscr{T}$ is in $\mathscr{T}$,

then the pair $(X, \mathscr{T})$ is called a *topological space*, and the members of $\mathscr{T}$ are called the *open sets* of $(X, \mathscr{T})$.

The study of topological spaces will not be pursued further here.

With the aid of Theorem 2.7.2, the following results can be deduced.

THEOREM 2.7.3. *Let (X, ρ) be a metric space;*

(i) *if $Y \subset Z \subseteq X$, then $\bar{Y} \subseteq \bar{Z}$;*

(ii) *if $Y \subseteq X$, then $\bar{Y}$ is the least closed set which contains Y (that is, if Z is any closed set such that $Y \subseteq Z$ then $\bar{Y} \subseteq Z$);*

(iii) *if $Y, Z \subseteq X$, then*

$$\overline{Y \cup Z} = \bar{Y} \cup \bar{Z},$$

but

$$\overline{Y \cap Z} \subseteq \bar{Y} \cap \bar{Z}.$$

Proof. (i) Let x be a limit point of Y and S be any neighbourhood of x. Then $(S-\{x\}) \cap Y \neq \varnothing$ and since $Y \subset Z, (S-\{x\}) \cap Z \neq \varnothing$, so $x \in \bar{Z}$; hence $\bar{Y} \subseteq \bar{Z}$.

(ii) This result follows immediately from (i).

(iii) Since $Y \subseteq \bar{Y}$ and $Z \subseteq \bar{Z}$, it follows that $Y \cup Z \subseteq \bar{Y} \cup \bar{Z}$; hence, from (i)

$$\overline{Y \cup Z} \subseteq \overline{\bar{Y} \cup \bar{Z}} = \bar{Y} \cup \bar{Z}$$

since $\bar{Y} \cup \bar{Z}$ is closed. Thus $\overline{Y \cup Z} \subseteq \bar{Y} \cup \bar{Z}$.

For the reverse inclusion, since $Y \subseteq Y \cup Z$, $Z \subseteq Y \cup Z$ it follows that

$$\bar{Y} \subseteq \overline{Y \cup Z}, \quad \bar{Z} \subseteq \overline{Y \cup Z}$$

so that $\bar{Y} \cup \bar{Z} \subseteq \overline{Y \cup Z}$.

For the second result observe that $Y \cap Z \subseteq Y$, $Y \cap Z \subseteq Z$, so that

$$\overline{Y \cap Z} \subseteq \bar{Y}, \quad \overline{Y \cap Z} \subseteq \bar{Z}$$

and hence $\overline{Y \cap Z} \subseteq \bar{Y} \cap \bar{Z}$.

It is left to the reader to construct an example to show that the case of strict inclusion is possible.

Exercises 2.7

1. (i) Let $\{Y_\lambda : \lambda \in \Lambda\}$ be a collection of closed sets in a metric space (X, ρ) and let Y denote the intersection of all the sets Y_λ. If x is any limit point of Y, show that it is also a limit point of Y_λ for all λ in Λ; deduce that Y is closed.

(ii) Let Y_1, Y_2 be two closed sets in (X, ρ). If x is a limit point of $Y_1 \cup Y_2$, show that it is also a limit point of at least one of Y_1, Y_2; deduce that $Y_1 \cup Y_2$ is closed. Hence prove that the union of any finite number of closed sets is closed.

2. Let (X, ρ) be a metric space and $Y \subseteq X$. Prove that the intersection of all the closed sets which contain Y is $\bar{Y}$.

3. If $\{Y_\lambda : \lambda \in \Lambda\}$ is any collection of sets in a metric space (X, ρ), prove that

$$\overline{\bigcup_{\lambda \in \Lambda} Y_\lambda} \supseteq \bigcup_{\lambda \in \Lambda} \bar{Y}_\lambda;$$

give an example to show that strict inclusion may occur when Λ is infinite (but not when Λ is finite).

2.8 Relative openness

It was explained in §2.1 that if (X, ρ) is a metric space and $Y \subset X$, then (Y, ρ_Y) is also a metric space. Despite this simple fact, we now come to the rude awakening that, if $Z \subset Y \subset X$, metric results concerning Z regarded as a subset of (Y, ρ_Y), do not necessarily hold when Z is regarded as a subset of (X, ρ) and, conversely, metric results concerning Z regarded as a subset of (X, ρ), do not necessarily hold when Z is regarded as a subset of (Y, ρ_Y). By a metric result we mean any result concerning metric spaces.

In the present section we shall consider an example of this; it is as follows. Let $Z \subseteq Y \subset X$, and ρ be a metric defined on X. If Z is open in (Y, ρ_Y), does it follow that Z is open in (X, ρ)? If Z is open in (X, ρ), does it follow that Z is open in (Y, ρ_Y)? The answers to these questions are 'not necessarily' and 'yes' respectively.

We give two examples of a subset of Y which is open in (Y, ρ_Y) but which is not open in (X, ρ). For the first example let (X, ρ) be any space having a subset Y which is not open in (X, ρ); then Y is open in (Y, ρ_Y), but not in (X, ρ).

For the second example, we consider $(\mathsf{R}^2, \mathsf{d})$. If $x_0 \in \mathsf{R}^2$, then any neighbourhood of x_0, regarded as a point of $(\mathsf{R}^2, \mathsf{d})$, is the interior of a circle, centre x_0. Consider next the subset $Y = \{(x, 0) : x \in \mathsf{R}\}$. Then (Y, d_Y) is a metric space; if $x_0 \in Y$, then a neighbourhood of x_0, regarded as a point of (Y, d_Y) is an interval, with mid-point x_0.

Now consider the set $Z = \{(x, 0) : a < x < b\}$ so that $Z \subset Y \subset X$. Then any point x_0 in Z is an interior point of Z when Z is considered as a subset of (Y, d_Y), since a neighbourhood of x_0 is then an interval, centre x_0. However x_0 is *not* an interior point of Z considered as a subset of $(\mathsf{R}^2, \mathsf{d})$, since any neighbourhood of x_0 is now the interior of a circle, which cannot possibly be contained in Z.

Thus, in this example $Z \subset Y \subset X$, and Z is open in (Y, d_Y), but is not open in (X, d).

The above example illustrates the important point that the nature of a neighbourhood of a point depends on the metric space which contains it. In connection with this we have Lemma 2.8.1.

If (X, ρ) is a metric space, $Y \subseteq X$ and $z \in Y$ then, for the remainder of this section, we denote the neighbourhoods of z (with radius r) in (Y, ρ_Y), (X, ρ) by $S_Y(z, r)$, $S_X(z, r)$ respectively.

LEMMA 2.8.1. *Let* (X, ρ) *be a metric space,* $Y \subseteq X$, $z \in Y$; *then*

$$S_Y(z, r) = Y \cap S_X(z, r).$$

Proof. This is left as an easy exercise.

Next a positive result concerning relative openness is given.

THEOREM 2.8.1. *Let* (X, ρ) *be a metric space,* $Z \subseteq Y \subseteq X$ *and* Z *be open in* (X, ρ); *then* Z *is open in* (Y, ρ_Y).

Proof. Let $z \in Z$; then there exists $\varepsilon > 0$ such that $S_X(z, \varepsilon) \subseteq Z$. Thus

$$Y \cap S_X(z, \varepsilon) \subseteq Z;$$

hence $S_Y(z, \varepsilon) \subseteq Z$, so z is an interior point of Z regarded as a subset of (Y, ρ_Y). Therefore Z is open in (Y, ρ_Y).

The next theorem contains the basic result concerning relative openness.

THEOREM 2.8.2. *Let* (X, ρ) *be a metric space, and* $Z \subseteq Y \subseteq X$; *then* Z *is open in* (Y, ρ_Y) *if and only if there exists a set* G *open in* (X, ρ) *such that* $Z = Y \cap G$.

Proof. First suppose that Z is open in (Y, ρ_Y). Then for each z in Z, there exists $\varepsilon > 0$ (where ε is dependent on z) such that $S_Y(z, \varepsilon) \subseteq Z$. Define

$$G = \bigcup_{z \in Z} S_X(z, \varepsilon);$$

then G is open in (X, ρ). Hence

$$Y \cap G = Y \cap \left\{ \bigcup_{z \in Z} S_X(z, \varepsilon) \right\} = \bigcup_{z \in Z} \{Y \cap S_X(z, \varepsilon)\} = \bigcup_{z \in Z} S_Y(z, \varepsilon),$$

using (1.1.2). But, as can be easily verified,

$$\bigcup_{z \in Z} S_Y(z, \varepsilon) = Z,$$

so there exists a set G open in (X, ρ) such that $Z = Y \cap G$.

Now assume that there exists a set G which is open in (X, ρ) and such that $Z = Y \cap G$. If $z \in Z$, then $z \in G$, so there exists $\varepsilon > 0$ such that $S_X(z, \varepsilon) \subseteq G$. Hence

$$Y \cap S_X(z, \varepsilon) \subseteq Y \cap G = Z,$$

that is $S_Y(z, \varepsilon) \subseteq Z$, so z is an interior point of Z in (Y, ρ_Y). Thus Z is open in (Y, ρ_Y).

COROLLARY 2.8.1. *Let (X, ρ) be a metric space and $Z \subseteq Y \subseteq X$; then Z is closed in (Y, ρ_Y) if and only if there exists a set H closed in (X, ρ) such that $Z = Y \cap H$.*

Proof. This result follows easily from the main theorem; the proof is left as an exercise.

Lastly we give a result which is analogous to Lemma 2.8.1; again we omit the proof, which is straightforward.

LEMMA 2.8.2. *Let (X, ρ) be a metric space and $Z \subseteq Y \subseteq X$. If $\bar{Z}^Y$, $\bar{Z}^X$ denote the closures of Z in (Y, ρ_Y), (X, ρ) respectively, then*

$$\bar{Z}^Y = Y \cap \bar{Z}^X.$$

EXERCISES 2.8

1. Let $X = [0, 1]$; which of the subsets $(\frac{1}{2}, 1)$, $(\frac{1}{2}, 1]$, $[\frac{1}{2}, 1)$ are open in (X, d)?

Which of these subsets are open in (i) (R, d), (ii) $(\mathsf{R}^2, \mathsf{d})$ where, in each case, d is the relevant Euclidean metric?

Which subsets of R are open in $(\mathsf{R}^2, \mathsf{d})$?

(For the purpose of this question identify R with the subset $\{(x, 0) : x \in \mathsf{R}\}$ of R^2.)

2. Let (X, ρ) be a metric space and $Y \subset X$. Show that every subset Z of Y, which is open in (Y, ρ_Y), will be open in (X, ρ) if and only if Y is open in (X, ρ).

2.9 Cluster values of a sequence

If (X, ρ) is a metric space and $Y \subseteq X$, then, by Theorem 2.5.1, the point x_0 is a limit point of Y if and only if there exists an infinite sequence (x_n) of distinct points all of which lie in Y, and such that $x_n \to x_0$ as $n \to \infty$. An analogous term is now defined for infinite sequences.

DEFINITION 2.9.1. Let (X, ρ) be a metric space, and (x_n) be an infinite sequence in X. Then the point c in X is called a *cluster point* (or *cluster value*) of the sequence if there exists a subsequence (x_{n_k}) which converges to c as $k \to \infty$.

Note that it is not required that all the points of the subsequence be distinct. The terms 'point of accumulation' and 'limit point' are also used by some authors instead of cluster point.

THEOREM 2.9.1. *Let (X, ρ) be a metric space and (x_n) be an infinite sequence in X; let $c \in X$. Then the following statements are equivalent:*

(i) *c is a cluster point of the sequence;*

(ii) *given any $\varepsilon > 0$, and any integer m, there exists $n > m$ such that $\rho(x_n, c) < \varepsilon$.*

Proof. Assume (i). Then there exists a subsequence (x_{n_k}) which converges to c as $k \to \infty$, so given any $\varepsilon > 0$, there exists K such that $\rho(x_{n_k}, c) < \varepsilon$ for all $k \geqslant K$. Thus taking any $K' \geqslant K$ such that $n_{K'} > m$ and letting $n = n_{K'}$, (ii) is established.

Now assume (ii), and define a subsequence as follows. Let n_1 be the least integer such that $\rho(x_{n_1}, c) < 1$. Define n_2 to be the least integer such that $n_2 > n_1$, and $\rho(x_{n_2}, c) < \frac{1}{2}$. Define n_3 to be the least integer such that $n_3 > n_2$, and $\rho(x_{n_3}, c) < \frac{1}{3}$, and so on (that is, by induction). The subsequence so obtained clearly converges to c; thus c is a cluster point of (x_n).

The reader familiar with the theory of upper and lower limits (in connection with real sequences) will know the following result.

THEOREM 2.9.2. *Let (x_n) be a real sequence and $C(x_n)$ denote the set of all cluster values of (x_n).*

If (x_n) is bounded above, then $C(x_n)$ is also bounded above; in this case, if $\Lambda = \sup C(x_n)$, then $\Lambda \in C(x_n)$. Similarly if (x_n) is bounded below, then $C(x_n)$ is also bounded below; in this case, if $\lambda = \inf C(x_n)$, then $\lambda \in C(x_n)$,

If (x_n) is bounded above and below, then (x_n) is convergent if and only if $\Lambda = \lambda$. If $\Lambda = \lambda$, then $x_n \to \Lambda$ as $n \to \infty$.

(The numbers Λ, λ are called the *upper* and *lower limits* of (x_n).)

EXERCISES 2.9

1. Find all the cluster values of the sequences

(i) $(\cos n\pi + (-1)^n n^{-1})$,

(ii) $(\frac{1}{2}a_n + 2^{-n})$ where $a_n \equiv n(\text{mod } 10)$,

in (R, d).

2. If (x_n) is a sequence in a metric space (X, ρ) having a cluster value c then show that, given any $\varepsilon > 0$, there exist an infinite number of values of n such that $\rho(x_n, c) < \varepsilon$.

If (x_{n_k}) is a subsequence (of (x_n)) which tends to c and which has an infinite number of distinct elements, then show that c is a limit point of the set whose members are the elements of the sequence.

2.10 Denseness

It is an elementary result concerning R that any real number can be expressed as the limit of a sequence of rationals; this can be demonstrated as follows.

Let $x \in \mathsf{R}$. Starting from the result that 'every open interval of R contains at least one rational, and hence an infinite number of rationals' (see Dieudonné (1960), (2.2.16)), we can construct a sequence of rationals converging to x by the following method. For by the result just quoted the interval $(x-n^{-1}, x+n^{-1})$ contains a rational q_n, say; then $q_n \to x$ as $n \to \infty$.†

Another way of expressing the same result is to say that for any x in R there exists a rational q which is arbitrarily close to x, that is, given any $\varepsilon > 0$, there exists a rational q such that $|x-q| < \varepsilon$. This leads us to say that 'the set of rationals is dense in the set of reals'. In view of Lemma 2.6.1 our assertion is equivalent to the statement that $\overline{\mathsf{Q}} = \mathsf{R}$ where closure is with respect to the Euclidean metric.

These comments motivate the next definition.

DEFINITION 2.10.1. Let (X, ρ) be a metric space, and $A \subseteq X$. Then A is said to be *everywhere dense* (in (X, ρ)) if $\bar{A} = X$.

LEMMA 2.10.1. *Let (X, ρ) be any metric space and $A \subseteq X$. Then the following statements are equivalent*:

(i) *A is everywhere dense*;

(ii) *for each x in X, there exists an infinite sequence (x_n) of points of A such that $x_n \to x$ as $n \to \infty$*;

(iii) *for each x in X, given any $\varepsilon > 0$ there exists x' in A such that $\rho(x, x') < \varepsilon$.*

Proof. This result follows immediately on using Lemma 2.6.1.

† The reader should convince himself that this argument can be refined to show that we can construct an increasing, or alternatively decreasing, sequence of rationals which converge to x. This result will be needed in §4.3.

EXERCISES 2.10

1. Let (X, ρ) be a metric space and $Y \subseteq X$. Prove that the following statements are equivalent:

(i) Y is everywhere dense in (X, ρ);
(ii) if Z is closed in (X, ρ) and $Y \subseteq Z \subseteq X$, then $Z = X$;
(iii) if Z' is open in (X, ρ) and $Y \cap Z' = \varnothing$, then $Z' = \varnothing$;
(iv) Y intersects every open sphere in (X, ρ).

2. Let Y_1, Y_2 be two sets both of which are open and everywhere dense in a metric space (X, ρ). Show that $Y_1 \cup Y_2$ and $Y_1 \cap Y_2$ are also open and everywhere dense.

2.11 The boundary, interior, exterior of a set

In connection with a subset of a metric space, so far we have defined the following basic terms: interior point, limit point, isolated point, open set, closed set, and the closure of a set. However there are other closely related terms which are sometimes of use; some of these are now described.

Given a simple figure in the real plane, such as a circular region, it is intuitively clear what is meant by the boundary of that figure. Our first definition generalizes this concept to any subset of any metric space.

DEFINITION 2.11.1. Let Y be a subset of a metric space (X, ρ); the set

$$\overline{Y} \cap \overline{(X-Y)}$$

is called the *boundary* of Y. It is denoted by $\partial(Y)$.

Thus the boundary of either of the subsets

$$\{(x_1, x_2) : x_1^2 + x_2^2 < 1\}, \quad \{(x_1, x_2) : x_1^2 + x_2^2 \leqslant 1\}$$

of $(\mathsf{R}^2, \mathsf{d})$ is the set

$$\{(x_1, x_2) : x_1^2 + x_2^2 = 1\}.$$

The boundary of the subset Q of (R, d) is R.

Of course the boundary of any set is closed.

DEFINITION 2.11.2. Let Y be a subset of a metric space (X, ρ); the set of all interior points of (X, ρ) is called the *interior* of Y. It is denoted by Y° or int Y.

For each y in $Y°$ there exists $\varepsilon_y > 0$ such that $S(y, \varepsilon_y) \subseteq Y$; then it is easily verified that

$$\bigcup_{y \in Y°} S(y, \varepsilon_y) = Y°.$$

It follows that $Y°$ is open. It is left to the reader to prove that $Y°$ is also equal to the union of all open sets contained in Y.

LEMMA 2.11.1. *Let Y be a subset of a metric space (X, ρ). Then*
(i) $Y° = X - \overline{X - Y}$;
(ii) $\bar{Y} = Y° \cup \partial(Y)$.

Proof. (i) First observe that $Y° \cap (\overline{X - Y}) = \varnothing$. For if $y \in Y°$ there exists $\varepsilon > 0$ such that $S(x, \varepsilon) \subseteq Y$ and so

$$S(x, \varepsilon) \cap (X - Y) = \varnothing;$$

thus $x \notin \overline{X - Y}$. Next suppose that x is any point of X. Then either there exists $\varepsilon > 0$ such that $S(x, \varepsilon) \subseteq Y$, when $x \in Y°$, or for all $\varepsilon > 0$

$$S(x, \varepsilon) \cap (X - Y) \neq \varnothing,$$

when $x \in \overline{X - Y}$. Thus (i) is established.

(ii) Since $Y° \subseteq Y \subseteq \bar{Y}$ and $\partial(Y) \subseteq \bar{Y}$ it follows that $Y° \cup \partial(Y) \subseteq \bar{Y}$. It remains to establish the reverse inclusion. Let $y \in \bar{Y}$; if $y \in Y°$ then $y \in Y° \cup \partial(Y)$. If $y \notin Y°$ then, for all $\varepsilon > 0$, $S(y, \varepsilon) \nsubseteq Y$, that is

$$S(y, \varepsilon) \cap (X - Y) \neq \varnothing;$$

thus $y \in \overline{X - Y}$, so $y \in \partial(Y)$ and hence $y \in Y° \cup \partial(Y)$. Therefore $\bar{Y} \subseteq Y° \cup \partial(Y)$.

DEFINITION 2.11.3. Let Y be a subset of a metric space (X, ρ); the interior of $X - Y$ is called the *exterior* of Y; it is denoted by ext Y. Any point of ext Y is said to be *exterior* to Y.

It is readily shown that, for any subset Y of (X, ρ),

$$X = \operatorname{int} Y \cup \partial(Y) \cup \operatorname{ext} Y,$$

where the sets on the right are disjoint, and also that

$$\operatorname{ext} Y = X - \bar{Y}.$$

EXERCISES 2.11

1. If Y is a subset of a discrete metric space (X, ρ), find the interior, exterior and boundary of Y.

2. If Y is a subset of a metric space (X, ρ) prove that

(i) $\partial(Y) \subseteq Y$ if and only if Y is closed;

(ii) $\partial(Y) \cap Y = \varnothing$ if and only if Y is open;

(iii) $\partial(Y) = \varnothing$ if and only if Y is both open and closed;

(iv) Y° is the largest open subset of Y, that is, if Z is any open set which is a subset of Y then $Z \subseteq Y^\circ$;

(v) Y is open if and only if $Y = Y^\circ$.

3. Let (X, ρ) be a metric space; prove the following results.

(i) If $Y \subset Z \subseteq X$ then $Y^\circ \subseteq Z^\circ$.

(ii) If $Y, Z \subset X$ then

$$(Y \cap Z)^\circ = Y^\circ \cap Z^\circ, \quad (Y \cup Z)^\circ \supseteq Y^\circ \cup Z^\circ.$$

Give an example for which the inclusion in the second result is strict.

(iii) For any $Y \subseteq X$, $\partial(Y^\circ) \subseteq \partial(Y)$.

4. If Y is an open subset of a metric space (X, ρ), show that

$$\text{int } Y \cup \text{ext } Y$$

is everywhere dense.

Show that the result also holds if Y is alternatively assumed to be closed, but that the result does not hold if no assumption is made concerning Y.

5. In the notation of Exercise 2.2.2, let $Y_i \subseteq X_i$ for each i. Show that

(i) $(Y_1 \times \ldots \times Y_m)^\circ = Y_1^\circ \times \ldots \times Y_m^\circ$,

(ii) $\overline{Y_1 \times \ldots \times Y_m} = \overline{Y}_1 \times \ldots \times \overline{Y}_m$,

where Y_i° is the interior of Y_i in (X_i, ρ_i) and $(Y_1 \times \ldots \times Y_m)^\circ$ is the interior of $Y_1 \times \ldots \times Y_m$ in (X, ρ) or (X, ρ'), and similar comments apply to the closures.

Deduce that if $Y_1, \ldots, Y_m$ are all non-empty then $Y_1 \times \ldots \times Y_m$ is open (closed) in (X, ρ), or in (X, ρ'), if and only if Y_i is open (closed) in (X_i, ρ_i) for each i.

3: CONTINUOUS FUNCTIONS

3.1 Introduction

As in real variable theory we shall be particularly interested in those functions which are continuous; the definition of such functions is a direct extension of the corresponding real variable definition.

DEFINITION 3.1.1. Let (X, ρ), (X', ρ') be two metric spaces; let f be a function of X into X'. Then

(i) f is said to be (ρ, ρ')-*continuous at* x_0 (in X) if, given any $\varepsilon > 0$, there exists $\delta > 0$ for which $\rho'(f(x), f(x_0)) < \varepsilon$ for all x such that $\rho(x, x_0) < \delta$;

(ii) f is said to be (ρ, ρ')-*continuous over* X if it is (ρ, ρ')-continuous at each point of X;

(iii) if furthermore $Y \subset X$, f is said to be (ρ, ρ')-*continuous over* Y if, for each x_0 in Y, given any $\varepsilon > 0$, there exists $\delta > 0$ for which $\rho'(f(x), f(x_0)) < \varepsilon$ for all x in Y such that $\rho(x, x_0) < \delta$.

We could replace (iii) by the following equivalent statement:

(iii)′ if furthermore $Y \subset X$, f is said to be (ρ, ρ')-*continuous over* Y if f_Y is (ρ_Y, ρ')-continuous over Y in the sense of (ii), where f_Y is the restriction of f to Y.

When discussing the continuity of f it is necessary to specify the metrics ρ, ρ' since it is possible that f may be continuous for a given pair of metrics ρ_1, ρ_1' defined on X, X' respectively, but may not be continuous for another pair of metrics, say ρ_2, ρ_2'; that is, f may be (ρ_1, ρ_1')-continuous while not being (ρ_2, ρ_2')-continuous. However this cumbersome notation can be suppressed by the following mathematically (somewhat imprecise) procedure, which we shall adopt.

If (X, ρ), (X', ρ') are two metric spaces and f is a function of X into X', then according to the definition of a function, f is entirely independent of the metric functions ρ, ρ'. Suppose now, however, that we talk of a function of (X, ρ) into (X', ρ'), so that the two metrics ρ, ρ' are associated with the name f of the function of X

into X' (we shall write $f:(X, \rho) \to (X', \rho')$); then a change in one or other metric implies a change in the identity of f. This enables us to make the following variation of Definition 3.1.1.

DEFINITION 3.1.1′. Let (X, ρ), (X', ρ') be two metric spaces; let f be a function of (X, ρ) into (X', ρ'). Then

(i) f is said to be *continuous at* x_0 (in X) if, given any $\varepsilon > 0$, there exists $\delta > 0$ for which $\rho'(f(x), f(x_0)) < \varepsilon$ for all x such that $\rho(x, x_0) < \delta$.

Similar modifications apply to the remainder of Definition 3.1.1.

The important point is that the metrics ρ, ρ' must be specified somewhere in the definition of continuity of a function between the sets of two metric spaces; the specification may be explicit (as in Definition 3.1.1) or implicit (as in Definition 3.1.1′).

The reader will observe that we have not considered the case of a function f whose domain of definition is a proper subset Y of (X, ρ). It is, of course, possible to vary the wording of the definition to cope with this situation; alternatively, and this will be our standpoint, we can regard f as a function from (Y, ρ_Y) into (X', ρ') and then use the above definitions.

The following results are immediate extensions of ones occurring in real Euclidean space.

THEOREM 3.1.1. *Let (X, ρ), (X', ρ') be two metric spaces, and f be a function of (X, ρ) into (X', ρ'); let $x_0 \in X$. Then the following statements are equivalent*:

(i) *f is continuous at x_0*;

(ii) *for every sequence (x_n) which converges to x_0 in (X, ρ), the corresponding sequence $(f(x_n))$ converges to $f(x_0)$ in (X', ρ').*

Proof. Assume (i); then (ii) follows easily.

Now assume (ii). Suppose, if possible, that f is not continuous at x_0; then by Lemma 1.7.2 there exists $\varepsilon_0 > 0$ such that for each $\delta > 0$ there is a point x for which $\rho(x, x_0) < \delta$ but $\rho'(f(x), f(x_0)) \geqslant \varepsilon_0$.

Taking $\delta = n^{-1}$, it follows that there exists x_n for which $\rho(x_n, x_0) < n^{-1}$ but $\rho'(f(x_n), f(x_0)) \geqslant \varepsilon_0$. Then $x_n \to x_0$, but $f(x_n)$ does not tend to $f(x_0)$. This gives a contradiction, so f is continuous at x_0, and the theorem is established.

THEOREM 3.1.2. *Let (X, ρ), (X', ρ'), (X'', ρ'') be metric spaces; if f, g are functions of (X, ρ) into (X', ρ') and (X', ρ') into (X'', ρ''), respectively, such that f is continuous at x_0 (in X) and g is continuous at $x_0' = f(x_0)$, then the function $g \circ f$ of (X, ρ) into (X'', ρ'') is continuous at x_0. Moreover if f, g are continuous over X, X' respectively then $g \circ f$ is continuous over X.*

Proof. This follows closely that of the corresponding Euclidean result; the details are omitted.

Lastly uniform continuity is defined; again this is a simple extension of the real variable term.

DEFINITION 3.1.2. Let (X, ρ), (X', ρ') be two metric spaces. A function f of (X, ρ) into (X', ρ') is said to be *uniformly continuous over* X if, for any $\varepsilon > 0$, there exists $\delta > 0$ for which $\rho'(f(x), f(y)) < \varepsilon$ for all x, y such that $\rho(x, y) < \delta$, where δ is independent of x, y.

If a function $f:(X, \rho) \to (X', \rho')$ is continuous (or uniformly continuous) over X, when no confusion can arise we shall often omit the words 'over X'.

EXERCISES 3.1

1. Let (X, ρ) be a metric space and $a \in X$; show that the function $f:(X, \rho) \to (\mathsf{R}, \mathsf{d})$ defined by $f(x) = \rho(x, a)$ is uniformly continuous over X.

2. Let the functions $f_n:(X, \rho) \to (Y, \sigma)$ be continuous for $n = 1, 2, \ldots$, and let the sequence (f_n) converge in (Y, σ) uniformly over X to $f:(X, \rho) \to (Y, \sigma)$. Prove that f is also continuous.
If, furthermore, each f_n is uniformly continuous over X show that f is also uniformly continuous over X.

3. Let $f:(X, \rho) \to (X', \rho')$ be continuous and $Y \subseteq X$. If y is a limit point of Y, is $f(y)$ necessarily a limit point of $f(Y)$? [Compare Exercise 3.3.3.]

4. Let (X, ρ), (X', ρ') be two metric spaces and f, g be two continuous mappings of (X, ρ) into (X', ρ'). Prove the following assertions.

(i) If $f(x) = g(x)$ for all x in A, then $f(x) = g(x)$ for all x in $\bar{A}$; thus if $B = \{x : f(x) = g(x)\}$, then B is closed.

(ii) If $X' = \mathsf{R}$, ρ' is the Euclidean metric and $C = \{x : f(x) \geqslant g(x)\}$, then C is closed.

3.2 Topological characterizations of continuity

The definition of continuity is essentially a statement concerning neighbourhoods. For, observe that the mapping $f:(X, \rho) \to (X', \rho')$ is continuous at x_0 (in X) if and only if, given any $\varepsilon > 0$, there exists $\delta > 0$ for which

$$\rho'(f(x), f(x_0)) < \varepsilon \quad \text{for all } x \text{ such that } \rho(x, x_0) < \delta,$$

that is

$$x \in S(x_0, \delta) \quad \text{implies that} \quad f(x) \in S'(f(x_0), \varepsilon),$$

or

$$f(S(x_0, \delta)) \subseteq S'(f(x_0), \varepsilon), \tag{3.2.1}$$

where S, S' denote spheres in (X, ρ), (X', ρ') respectively. This can be rewritten in a slightly different form. To see this, if $N \subseteq X$, $N' \subseteq X'$ using (1.2.3), (1.2.14), (1.2.8), (1.2.13) it follows that $f(N) \subseteq N'$ if and only if $N \subseteq f^{-1}(N')$; hence (3.2.1) can be replaced equivalently by

$$S(x_0, \delta) \subseteq f^{-1}(S'(f(x_0), \varepsilon)). \tag{3.2.2}$$

Thus we have proved the following result.

LEMMA 3.2.1. *Let f be a mapping of (X, ρ) into (X', ρ'). Then f is continuous at x_0 (in X) if and only if, given any $\varepsilon > 0$, there exists $\delta > 0$ such that* (3.2.1), *or equivalently* (3.2.2), *holds.*

Next an important result concerning continuity over a metric space will be developed.

THEOREM 3.2.1. *Let f be a mapping of (X, ρ) into (X', ρ'). Then the following statements are equivalent:*

(i) *f is continuous over X;*

(ii) *for each set Y' open in (X', ρ'), the set $f^{-1}(Y')$ is open in (X, ρ);*

(iii) *for each set Y' closed in (X', ρ'), the set $f^{-1}(Y')$ is closed in (X, ρ);*

(iv) *for each set $Y \subseteq X$, $f(\bar{Y}) \subseteq \overline{f(Y)}$.*

Proof. Assume (i); let Y' be any open set in (X', ρ'). Let $x_0 \in f^{-1}(Y')$, so $f(x_0) \in Y'$; then there exists $\varepsilon > 0$ such that $S'(f(x_0), \varepsilon) \subseteq Y'$. Since f is continuous at x_0, by Lemma 3.2.1, for this $\varepsilon > 0$ there exists $\delta > 0$ such that

$$S(x_0, \delta) \subseteq f^{-1}(S'(f(x_0), \varepsilon)) \subseteq f^{-1}(Y'),$$

using (1.2.8). Thus x_0 is an interior point of $f^{-1}(Y')$, so $f^{-1}(Y')$ is open in (X, ρ).

Now assume (ii); we shall deduce (i). Let $x_0 \in X$; then given any $\varepsilon > 0$, the neighbourhood $S'(f(x_0), \varepsilon)$ is open in (X', ρ'), so by hypothesis $f^{-1}(S'(f(x_0), \varepsilon))$ is open in (X, ρ). Since

$$x_0 \in f^{-1}(S'(f(x_0), \varepsilon))$$

it follows that there exists $\delta > 0$ such that

$$S(x_0, \delta) \subseteq f^{-1}(S'(f(x_0), \varepsilon)).$$

Hence, by Lemma 3.2.1, f is continuous at x_0 and so is continuous over X.

So far it has been shown that (i) and (ii) are equivalent.

Assume, again, (ii) and let Y' be a closed set in (X', ρ'). Then $X' - Y'$ is open in (X', ρ') so $f^{-1}(X' - Y')$ is open in (X, ρ). But

$$f^{-1}(X' - Y') = X - f^{-1}(Y'),$$

by (1.2.15), so $f^{-1}(Y')$ is closed in (X, ρ); this establishes (iii). In the same way it is seen that (iii) implies (ii).

Finally (iii) and (iv) will be shown to be equivalent.

Assume (iii), and let $Y \subseteq X$, then $f^{-1}(\overline{f(Y)})$ is closed in (X, ρ). But

$$Y \subseteq f^{-1}(f(Y)) \subseteq f^{-1}(\overline{f(Y)});$$

$\bar{Y}$ is the least closed set containing Y (by Theorem 2.7.3), so $\bar{Y} \subseteq f^{-1}(\overline{f(Y)})$. Hence

$$f(\bar{Y}) \subseteq f(f^{-1}(\overline{f(Y)})) \subseteq \overline{f(Y)}.$$

Now assume (iv); let Y' be a set closed in (X', ρ') and set $Y = f^{-1}(Y')$. Then

$$(f)\, \bar{Y} \subseteq \overline{f(f^{-1}(Y'))} \subseteq \bar{Y}' = Y',$$

so that

$$\bar{Y} \subseteq f^{-1}(f(\bar{Y})) \subseteq f^{-1}(Y') = Y.$$

Thus $Y = \bar{Y}$, so Y is closed, that is $f^{-1}(Y')$ is closed (in (X, ρ)).

In conclusion we make two remarks.

(*a*) The equivalence of (i) and (ii) is of fundamental significance; for this reason it was proved that (i) implies (ii) and (ii) implies (i). The equivalence of (ii) and (iii) is elementary. We have therefore not proved the equivalence of the results by the quickest route, but rather have used one which is intended to stress the inter-relation of the four statements.

The equivalence of (i) and (ii) is important because it enables us to describe continuity in terms of open sets instead of relations involving metrics; this is of importance when studying continuity in a topological space in the absence of a metric.

(*b*) In the characterizations (ii), (iii) of continuity it should be noted that it is the *inverse* image of an open (or closed) set which is an open (or closed) set. It is not in general true that the direct image of an open set is open, or of a closed set is closed. For example consider the mapping $f:(\mathsf{R}, \mathsf{d}) \to (\mathsf{R}, \mathsf{d})$ where

$$f(x) = \frac{x^2}{1+x^2};$$

then $f(\mathsf{R}) = [0, 1)$ which is neither open nor closed, although R is both open and closed.

EXERCISES 3.2

1. Let (X, ρ), (X, ρ') be two metric spaces having the same underlying set X; let $\mathscr{T}, \mathscr{T}'$ denote the collections of all open sets of (X, ρ), (X, ρ') respectively. Let $i:(X, \rho) \to (X, \rho')$ be the identity mapping (see §1.2); show that i is continuous if and only if $\mathscr{T}' \subseteq \mathscr{T}$.

2. Let (X, ρ) be a discrete space and (Y, σ) be any metric space; let f be any mapping of (X, ρ) into (Y, σ). Show that f is continuous.

3. Give an example of a function $f:(X, \rho) \to (X', \rho')$ which is not continuous and, for this function, find

(i) a set Y' which is open in (X', ρ') but such that $f^{-1}(Y')$ is not open in (X, ρ);

(ii) a set Z' which is closed in (X', ρ') but such that $f^{-1}(Z')$ is not closed in (X, ρ);

(iii) a set $Y \subseteq X$ such that $f(\bar{Y}) \nsubseteq \overline{f(Y)}$.

4. Let $f:(X, \rho) \to (X', \rho')$ be continuous over X and let $Y \subset X$. Show that the restriction f_Y is continuous over Y by using the epsilon-delta definition of continuity and, alternatively, the open set characterization of continuity.

5. Let $f:(X, \rho) \to (X', \rho')$ be continuous and $Y \subset X$. If $y \in Y$ and $f(y)$ is an interior point of $f(Y)$, prove that y is an interior point of $f^{-1}(f(Y))$.

By means of the following example show that y is not, however, necessarily an interior point of Y. Let $X = X' = \mathsf{R}$, $\rho = \rho'$ be the Euclidean metric, and Y be the union of the rationals in $[0, 1]$ and the irrationals in $(-1, 0)$; let f be defined by $f(x) = |x|$.

6. Let (X, ρ), (X', ρ') be two metric spaces and let Y, Z be subsets of X such that $Y \cup Z = X$. Let the functions $f:(Y, \rho_Y) \to (X', \rho')$, $g:(Z, \rho_Z) \to (X', \rho')$ be continuous, and $f(x) = g(x)$ for all x in $Y \cap Z$. Define $h:(X, \rho) \to (X', \rho')$ by

$$h(x) = \begin{cases} f(x) & \text{for } x \in Y, \\ g(x) & \text{for } x \in Z; \end{cases}$$

show that h is not necessarily continuous.

If, however, it is also assumed that Y and Z are either both open, or are both closed, in (X, ρ), then with the aid of Exercise 2.8.2, or otherwise, prove that h must be continuous.

7. Let f be a mapping of (X, ρ) into (X', ρ'). Prove that the following statements are equivalent:

(i) f is continuous over X;

(ii) for each set $Y' \subseteq X'$, $\overline{f^{-1}(Y')} \subseteq f^{-1}(\bar{Y}')$;

(iii) for each set $Y' \subseteq X'$, $\{f^{-1}(Y')\}^{\circ} \supseteq f^{-1}(Y'^{\circ})$.

8. In the notation of Exercise 2.2.2 let the mapping $\pi_i : X \to X_i$ be defined by

$$\pi_i(x_1, \ldots, x_n) = x_i;$$

show that π_i is a uniformly continuous mapping of (X, ρ), or (X, ρ'), onto (X_i, ρ_i). (The function π_i is known as the *projection* of X onto X_i.)

Extend this result to the metric space defined in Exercise 2.2.4.

3.3 Homeomorphisms

If f is a bijective mapping of X onto X', then it is elementary that f possesses an inverse (denoted by f^{-1}), and that f^{-1} is a bijective mapping of X' onto X. Suppose now that (X, ρ) and (X', ρ') are metric spaces; if, furthermore, $f:(X, \rho) \to (X', \rho')$ is continuous, it does not necessarily follow that $f^{-1}:(X', \rho') \to (X, \rho)$ is also continuous. An example to illustrate this will be given shortly; first we state a definition motivated by these comments.

DEFINITION 3.3.1. Let (X, ρ), (X', ρ') be two metric spaces and $f:(X, \rho) \to (X', \rho')$ be a bijection of X onto X' such that f is continuous over X and f^{-1} is continuous over X'; then f is said to be a *homeomorphism*.†

If $f:(X, \rho) \to (X', \rho')$ is a homeomorphism, it follows immediately that $f^{-1}:(X', \rho') \to (X, \rho)$ is also a homeomorphism.

Two metric spaces (X, ρ), (X', ρ') are said to be *homeomorphic* (or *homeomorphic images* of each other) if there exists a bijective mapping $f:(X, \rho) \to (X', \rho')$ which is a homeomorphism.

Now an example is given in which $f:(X, \rho) \to (X', \rho')$ is a continuous bijection but such that $f^{-1}:(X', \rho') \to (X, \rho)$ is not continuous. Let $X = X' = \mathsf{R}$, ρ be the standard discrete metric, ρ' be the Euclidean metric, and let f be the identity map on R. Then, if $x_0 \in X$, given any $\varepsilon > 0$ and taking any δ in $(0, 1)$, $\rho'(f(x), f(x_0)) < \varepsilon$ for all x such that $\rho(x, x_0) < \delta$ (since the only x to satisfy the second inequality is $x = x_0$, for which the first inequality is trivially satisfied), so f is continuous at x_0 and thus is continuous over X. However, given any $\varepsilon > 0$ such that $\varepsilon < 1$, there exists no $\delta > 0$ for which $\rho(f^{-1}(x), f^{-1}(x_0)) < \varepsilon$ for all x such that $\rho'(x, x_0) < \delta$; for, if $\delta > 0$ there exists $x \neq x_0$ such that $\rho'(x, x_0) < \delta$, and for this x

$$\rho(f^{-1}(x), f^{-1}(x_0)) = 1 \nless \varepsilon,$$

so f^{-1} is not continuous at any point x_0 of X'. (For future reference note, also, that f is uniformly continuous over X.)

Since a homeomorphism is a continuous mapping having a continuous inverse, Theorems 3.1.1 and 3.2.1 immediately imply the following result.

† Not to be confused with *homomorphism* which is an algebraic term describing mappings which preserve certain laws of composition.

THEOREM 3.3.1. *Let f be a bijective mapping of (X, ρ) onto (X', ρ'). Then the following statements are equivalent:*

(i) *f is a homeomorphism;*

(ii) *for each $Y \subseteq X$, Y is open in (X, ρ) if and only if $f(Y)$ is open in (X', ρ');*

(iii) *for each $Y \subseteq X$, Y is closed in (X, ρ) if and only if $f(Y)$ is closed in (X', ρ');*

(iv) *for each set $Y \subseteq X$, $f(\bar{Y}) = \overline{f(Y)}$, (where $\bar{Y}$ is the closure of Y in (X, ρ) and $\overline{f(Y)}$ is the closure of $f(Y)$ in (X', ρ'));*

(v) *for each sequence (x_n) which converges to x_0 in (X, ρ), $(f(x_n))$ converges to $f(x_0)$ in (X', ρ'), and for each sequence (x'_n) which converges to x'_0 in (X', ρ'), $(f^{-1}(x'_n))$ converges to $f^{-1}(x'_0)$ in (X, ρ).*

Here again the equivalence of (i) and (ii) is of fundamental significance. For, if (X, ρ), (X', ρ') are homeomorphic there is a bijection between the collections of open sets of (X, ρ) and of (X', ρ'), so if we have a result concerning a metric space (X, ρ) which involves only the concept of open sets (for example, the concept of continuity) we will immediately have a corresponding result for (X', ρ').

We make one final remark; the property of two metric spaces being homeomorphic is an equivalence relation. That is, if $\mathcal{M}$ is a collection of metric spaces and H is the relation (in $\mathcal{M}$) defined by EHF if and only if E is homeomorphic to F, where $E, F \in \mathcal{M}$, then H is an equivalence relation.

EXERCISES 3.3

1. Let $I = (0, 1)$; by finding a suitable bijection from I to R show that (R, d), (I, d) are homeomorphic. If I_1, I_2 are open intervals show that (I_1, d), (I_2, d) are homeomorphic. (Here d denotes the Euclidean metric on R or on any subset of R.)

2. (i) Let (X, ρ), (X', ρ') be homeomorphic. Show that (X, ρ) is discrete if and only if (X', ρ') is discrete.

(ii) Two sets X, X' are said to be *equipotent* if there exists a bijective mapping between them. If X, X' are equipotent and ρ, ρ' are discrete metrics associated with X, X', show that (X, ρ), (X', ρ') are homeomorphic.

3. Let $f:(X, \rho) \to (X', \rho')$ be a homeomorphism, $Y \subset X$ and $y \in Y$. Prove that

(i) y is a limit point of Y if and only if $f(y)$ is a limit point of $f(Y)$;

(ii) y is an interior point of Y if and only if $f(y)$ is an interior point of $f(Y)$. [Compare Exercise 3.2.5.]

4. Let $f:(X, \rho) \to (Y, \sigma)$ be a homeomorphism, and $Z \subset X$ be such that $Z \cap Z' = \varnothing$ where Z' denotes the set of all limit points of Z in (X, ρ). Show that $f(Z) \cap \{f(Z)\}' = \varnothing$.

5. Let X_1, X_2, X_3 be three sets and $f: X_1 \to X_2$, $g: X_2 \to X_3$ be two mappings such that $g \circ f$ is bijective. If either g is injective or f is surjective show that f, g are both bijective.

Now associate metrics ρ_i with the sets X_i $(i = 1, 2, 3)$ respectively. If moreover f, g are continuous and $g \circ f$ is a homeomorphism, prove that f, g are both homeomorphisms.

6. Let $X = [0, 2\pi)$ and let the function $f: X \to f(X) (\subseteq \mathsf{R}^2)$ be defined by $f(t) = (\cos t, \sin t)$ for all t in X. Associate the relevant Euclidean metrics with X and $f(X)$. Prove that f is a continuous bijection, but that f^{-1} is not continuous.

3.4 Equivalent metrics

In this section we look at the case of homeomorphic spaces having the same underlying set; a special term is used to describe this situation.

DEFINITION 3.4.1. Let (X, ρ), (X, ρ') be two metric spaces (having the same underlying set); if the identity mapping $i:(X, \rho) \to (X, \rho')$ is a homeomorphism, then the metrics ρ, ρ' are said to be *equivalent* on X.

Theorem 3.3.1 may now be restated in the following form.

THEOREM 3.4.1. *Let (X, ρ), (X, ρ') be two metric spaces. Then the following statements are equivalent*:

(i) *the metrics ρ, ρ' are equivalent*;

(ii) *for each $Y \subseteq X$, Y is open in (X, ρ) if and only if Y is open in (X, ρ')*;

(iii) *for each $Y \subseteq X$, Y is closed in (X, ρ) if and only if Y is closed in (X, ρ')*;

(iv) *for each* $Y \subseteq X$, $\bar{Y}^{\rho} = \bar{Y}^{\rho'}$, *where* $\bar{Y}^{\rho}$, $\bar{Y}^{\rho'}$ *denote the closures of* Y *in* (X, ρ), (X, ρ') *respectively*;

(v) *the sequence* (x_n) *converges to* x_0 *in* (X, ρ) *if and only if it converges to* x_0 *in* (X, ρ').

It is clear that the relation of two metrics being equivalent is indeed an equivalence relation.

The following result gives a sufficient condition for two metrics to be equivalent.

LEMMA 3.4.1. *Let* (X, ρ), (X, ρ') *be two metric spaces; if there exist strictly positive constants* α, β *such that*

$$\alpha\rho(x, y) \leqslant \rho'(x, y) \leqslant \beta\rho(x, y) \tag{3.4.1}$$

for all x, y *in* X, *then the metrics* ρ, ρ' *are equivalent.*

Proof. Given any $\varepsilon > 0$, let $\delta = \varepsilon/\beta$; since $\rho' \leqslant \beta\rho$ it follows that $\rho'(x, y) < \varepsilon$ for all x, y such that $\rho(x, y) < \delta$. Therefore the identity mapping of (X, ρ) into (X, ρ') is continuous over X; in fact it is uniformly continuous over X since δ is independent of x, y. Similarly it follows that the identity mapping of (X, ρ') into (X, ρ) is also uniformly continuous over X. Hence ρ, ρ' are equivalent over X.

If we agree to say that ρ, ρ' are *uniformly equivalent* when the identity mapping of (X, ρ) onto (X, ρ') and its inverse are both uniformly continuous over X, then the above proof shows that (3.4.1) implies that ρ, ρ' are uniformly equivalent. The relation of two metrics being uniformly equivalent is also an equivalence relation.

Example. Let $X = \mathsf{R}^m$, $x = (x_1, \ldots, x_m)$, $y = (y_1, \ldots, y_m)$,

$$\rho(x, y) = \left\{\sum_{i=1}^{m} |x_i - y_i|^p\right\}^{1/p}, \tag{3.4.2}$$

where $p \geqslant 1$, and

$$\rho'(x, y) = \max_{1 \leq i \leq m} |x_i - y_i|.$$

Then $\rho(x, y) \geqslant |x_i - y_i|$ for $i = 1, \ldots, m$ and so $\rho(x, y) \geqslant \rho'(x, y)$ for all x, y in X. Furthermore

$$\rho(x, y) \leqslant \{m(\max_i |x_i - y_i|)^p\}^{1/p},$$

so that $\rho(x, y) \leqslant m^{1/p}\rho'(x, y)$ for all x, y in X.

Thus by Lemma 3.4.1 ρ, ρ' are (uniformly) equivalent over X.

Since the relation of metrics being equivalent is an equivalence relation, it follows that the metrics ρ defined by (3.4.2), for different p, are all equivalent.

From the equivalence of the statements (i) and (v) in Theorem 3.4.1, it follows that $(x^{(n)})$ converges to x in (X, ρ') if and only if $(x^{(n)})$ converges to x in (X, ρ) for any $p \geqslant 1$ (including, of course, the Euclidean case of $p = 2$); this is precisely the conclusion we arrived at in §2.4 (ii)–(iv) by a direct consideration of convergence in each of the metric spaces.

Finally an example is given which shows the condition (3.4.1), although a sufficient condition for ρ, ρ' to be uniformly equivalent, is not a necessary one.

Let $X = \mathsf{R}$, $\rho(x, y) = |x-y|$, and

$$\rho'(x, y) = \frac{|x-y|}{1+|x-y|};$$

it is easily verified that (R, ρ') is a metric space (see Exercise 2.1.4). Since $\rho'(x, y) \leqslant \rho(x, y)$ for all x, y in R, it follows immediately (as in the first part of the proof of Lemma 3.4.1) that the identity mapping of (X, ρ) onto (X, ρ') is uniformly continuous. However it is clear that there exists no constant α such that $\alpha\rho(x, y) \leqslant \rho'(x, y)$ for all x, y in R; nevertheless the identity mapping of (X, ρ') onto (X, ρ) is uniformly continuous. To show this, observe that if $\varepsilon > 0$ and

$$\frac{|x-y|}{1+|x-y|} < \frac{\varepsilon}{1+\varepsilon}$$

then $|x-y| < \varepsilon$; hence given any $\varepsilon > 0$, if we set $\delta = \varepsilon/(1+\varepsilon)$, it follows that $\rho(x, y) < \varepsilon$ for all x, y such that $\rho'(x, y) < \delta$, where δ is independent of x, y.

As yet, the concept of boundedness, so important in the real Euclidean space, has not been generalized to arbitrary metric spaces; this is now done.

If a set Y of real numbers possesses the property that there exists M such that $|y| < M$ for all y in Y, then Y is said to be bounded. Since metric spaces do not possess an order relation we cannot generalize this definition of boundedness; therefore we look for a characterization of a bounded real set which can be generalized. This is easily found; the one we choose is as follows. The set Y of real numbers is bounded if and only if there exists an interval $(-M, M)$ such that $Y \subseteq (-M, M)$.

DEFINITION 3.4.2. A subset Y of a metric space (X, ρ) is said to be *bounded in* (X, ρ) if there exists a sphere S such that $Y \subseteq S$. If X is a bounded set in (X, ρ) then (X, ρ) is said to be a bounded metric space.

Observe that (X, ρ) is a bounded metric space if and only if the set $\{\rho(x, y) : x, y \in X\}$ of real numbers is bounded.

LEMMA 3.4.2. *For any metric space* (X, ρ) *there exists a metric* ρ' *uniformly equivalent to* ρ *such that* (X, ρ') *is bounded.*

Proof. The metric defined by

$$\rho'(x, y) = \frac{\rho(x, y)}{1+\rho(x, y)}$$

will suffice. The details are left to the reader.

EXERCISES 3.4

1. Let (X, ρ) be a metric space and define

$$\rho_1(x, y) = \frac{\rho(x, y)}{1+\rho(x, y)}, \quad \rho_2(x, y) = \min\{1, \rho(x, y)\}.$$

Show that (X, ρ_1), (X, ρ_2) are both bounded metric spaces and ρ_1, ρ_2 are both uniformly equivalent to ρ.

2. In the notation of Exercise 2.2.2, show that ρ, ρ' are uniformly equivalent.

3. Are either of the metrics defined in Exercises 2.1.5, 2.1.6 equivalent to the Euclidean metric on R^2?

4. Let X be a finite non-empty set and ρ, ρ' be any two metrics associated with X; show that ρ, ρ' are equivalent.

5. Associate with the set $\mathscr{C}(I)$, where $I = [a, b]$, the two metrics defined by

$$\rho(f, g) = \sup_t |f(t)-g(t)|, \quad \rho'(f, g) = \int_a^b |f(t)-g(t)|\, dt.$$

Show that ρ, ρ' are not equivalent.

6. (i) Show that every discrete metric (on a non-empty set X) is equivalent to the standard discrete metric (on X).

(ii) Define $\rho:\mathsf{N}\times\mathsf{N}\to\mathsf{R}$ by $\rho(m, n) = |m^{-1}-n^{-1}|$; show that (N, ρ) is a discrete metric space but that ρ is not uniformly equivalent to the standard discrete metric on N.

3.5 The distances $\rho(x, Y)$, $\rho(Y, Z)$

We now define the distance of a point from a set and the distance between two sets.

DEFINITION 3.5.1. Let (X, ρ) be a metric space, Y, Z be non-empty subsets of X and let $x \in X$. Then

$$\inf_{y \in Y} \rho(x, y)$$

is called the *distance of x from Y, and is denoted by* $\rho(x, Y)$; furthermore

$$\inf_{(y,z) \in Y \times Z} \rho(y, z)$$

is called the *distance between* Y, Z, and is denoted by $\rho(Y, Z)$.

LEMMA 3.5.1. *Let (X, ρ) be a metric space, Y be a non-empty subset of X and $x \in X$. Then the function $f:(X, \rho) \to (\mathsf{R}, \mathsf{d})$ defined by $f(x) = \rho(x, Y)$ is uniformly continuous over X.*

Proof. For any x_0 in X it must be shown that, given any $\varepsilon > 0$, there exists $\delta > 0$ for which

$$|\rho(x, Y)-\rho(x_0, Y)| < \varepsilon \quad \text{for all } x \text{ such that } \rho(x, x_0) < \delta, \qquad (3.5.1)$$

where δ is independent of x_0.

Now for all y in Y

$$\rho(x, Y) \leqslant \rho(x, y) \leqslant \rho(x, x_0)+\rho(x_0, y);$$

hence
$$\rho(x, Y) \leqslant \rho(x, x_0)+\rho(x_0, Y),$$

and so
$$\rho(x, Y)-\rho(x_0, Y) \leqslant \rho(x, x_0).$$

Similarly
$$\rho(x_0, Y)-\rho(x, Y) \leqslant \rho(x_0, x),$$

and thus
$$|\rho(x, Y)-\rho(x_0, Y)| \leqslant \rho(x, x_0);$$

hence taking $\delta = \varepsilon$ (so that δ is independent of x_0), (3.5.1) is satisfied. Thus f is uniformly continuous over X.

If $x \in Y$, then clearly $\rho(x, Y) = 0$. However the converse of this is not true; that is $\rho(x, Y) = 0$ does not necessarily imply that $x \in Y$ (for example if $X = \mathsf{R}$, $Y = (a, b)$ and $x = a$, then $\mathsf{d}(x, Y) = 0$). The complete answer to 'when does $\rho(x, Y) = 0$?' is given in the next result.

LEMMA 3.5.2. *Let (X, ρ) be a metric space, $x \in X$ and Y be a non-empty subset of X. Then $\rho(x, Y) = 0$ if and only if $x \in \bar{Y}$.*

Proof. By Lemma 2.6.1, $x \in \bar{Y}$ if and only if given any $\varepsilon > 0$, $S(x, \varepsilon) \cap Y \neq \varnothing$, that is $\rho(x, Y) < \varepsilon$. Thus $x \in \bar{Y}$ if and only if $\rho(x, Y) = 0$.

EXERCISES 3.5

1. Show that

$$\inf_{(y,z) \in Y \times Z} \rho(y, z) = \inf_{y \in Y} \left\{ \inf_{z \in Z} \rho(y, z) \right\},$$

that is

$$\rho(Y, Z) = \inf_{y \in Y} \rho(y, Z)$$

[This result holds for *any* function $\rho: Y \times Z \to \mathsf{R}$ which is bounded below.]

2. Let (X, ρ) be a metric space and $\{C_\lambda : \lambda \in \Lambda\}$ be a class of closed subsets with the property that there exists $\delta > 0$ such that $\rho(C_\lambda, C_{\lambda'}) \geqslant \delta$ for all $\lambda \neq \lambda'$. Prove that the union of all the sets C_λ is closed.

3. Let $A = \{2, 3, 4, \ldots\}$, $B = \{2\frac{1}{2}, 3\frac{1}{3}, 4\frac{1}{4}, \ldots\}$; show that A, B are two disjoint closed sets of (R, d) such that $\mathsf{d}(A, B) = 0$. [Thus if A, B are disjoint closed subsets of a metric space (X, ρ) it does not necessarily follow that $\rho(A, B) \neq 0$. See also (iii) of Exercise 5.8.5.]

4. Let (X, ρ) be a metric space and Y be a non-empty subset of X; then

$$\sup_{x,y \in Y} \rho(x, y)$$

is called the *diameter* of Y, and is denoted by $d(Y)$. Show that

(i) Y is bounded if and only if $d(Y)$ is finite;

(ii) $d(Y) = d(\bar{Y})$;

(iii) if Y, Z are non-empty subsets of X then

$$d(Y \cup Z) \leqslant d(Y) + d(Z) + \rho(Y, Z).$$

4: COMPLETENESS

4.1 Complete metric spaces

We now turn to some aspects of convergence in metric spaces. First the reader is reminded of the following result concerning real sequences.

PROPOSITION 4.1.1 (The Cauchy principle of convergence). *The real sequence* (x_n) *converges to a limit in* R *if and only if given any* $\varepsilon > 0$ *there exists* N *such that* $|x_m - x_n| < \varepsilon$ *for all* $m, n > N$.

Since this is a basic result concerning convergence in R, the question arises as to whether there is an analogue for general metric spaces; that is, in any metric space (X, ρ), can it be shown that a sequence (x_n) converges to a limit in X if and only if given any $\varepsilon > 0$ there exists N such that $\rho(x_m, x_n) < \varepsilon$ for all $m, n > N$? The answer is 'no'. For, in the metric space (Q, d) let the sequence (x_n) be defined by

$$x_n = 1+\frac{1}{1!}+\frac{1}{2!}+\frac{1}{3!}+\ldots+\frac{1}{n!};$$

then, as is well known, the sequence (x_n) does not tend to a limit in Q, despite that given any $\varepsilon > 0$ there exists N such that $|x_m - x_n| < \varepsilon$ for all $m, n > N$.

We make some definitions.

DEFINITION 4.1.1. A sequence (x_n) of points in a metric space (X, ρ) is called a *fundamental* (or *Cauchy*) sequence in (X, ρ) if, given any $\varepsilon > 0$, there exists N such that $\rho(x_m, x_n) < \varepsilon$ for all $m, n > N$ (that is, if $\rho(x_m, x_n) \to 0$ as $m, n \to \infty$).

Roughly speaking a fundamental sequence (x_n) is one for which the members x_n become closer and closer to each other as $n \to \infty$.

It follows that any sequence (x_n) in (X, ρ) which is convergent must be fundamental.

The question that was asked above can now be expressed as 'does every fundamental sequence of a metric space (X, ρ) converge to a limit (in X)?'. The answer, as we have seen by example, is 'no'. This leads us to study those metric spaces in which every fundamental sequence does converge to a limit in the metric space.

DEFINITION 4.1.2. The metric space (X, ρ) is said to be *complete* if every fundamental sequence in (X, ρ) is convergent in (X, ρ).

Thus, (Q, d) is not complete, but (R, d) is complete. Again roughly speaking, just as Q contains 'gaps' in the sense that there exist sequences in Q (such as the one described above) whose elements become closer and closer to each other without converging to an element of Q (but instead becoming closer and closer to a gap), in the same way metric spaces which are not complete possess gaps.

It will be remembered that if Y is a non-empty subset of X, then a point x_0 of X is a limit point of Y if and only if there exists a sequence of distinct points of Y which converges to x_0 (see Theorem 2.5.1); furthermore Y is closed if and only if it contains all its limit points (see Theorem 2.6.1). When discussing completeness (a concept defined in terms of convergence) these characterizations of limit points and closed sets will be particularly useful since they describe these concepts also in terms of convergence. This is illustrated in the proofs of the next two simple results concerning the completeness of subspaces.

THEOREM 4.1.1. *If Y is a non-empty subset of a complete metric space (X, ρ) then (Y, ρ_Y) is complete if and only if Y is closed in (X, ρ).*

Proof. First suppose that Y is closed in (X, ρ), and let (x_n) be a fundamental sequence of (Y, ρ_Y). Then (x_n) is also a fundamental sequence of (X, ρ), so there exists x_0 in X such that $x_n \to x_0$; but by Lemma 2.6.1, $x_0 \in \bar{Y}$ so $x_0 \in Y$. Hence (Y, ρ_Y) is complete.

Conversely suppose that (Y, ρ_Y) is complete. Let x be any limit point of Y, so there exists a sequence (x_n) in Y which converges to x; then (x_n) is a fundamental sequence of (Y, ρ_Y) and so its limit x belongs to Y. Hence Y is closed.

In the second half of the preceding proof we did not use that (X, ρ) is complete. Hence we have the following result.

THEOREM 4.1.2. *If Y is a non-empty subset of a metric space (X, ρ) and if (Y, ρ_Y) is complete, then Y is closed in (X, ρ).*

We establish two other results which will be required later.

THEOREM 4.1.3. *If (x_n) is a fundamental sequence of a metric space (X, ρ), and if (x_n) has a cluster value c in X, then the sequence converges (to c).*

Proof. Given any $\varepsilon > 0$ there exists N such that $\rho(x_m, x_n) < \frac{1}{2}\varepsilon$ for all $m, n > N$. Since c is a cluster value of the sequence, by Theorem 2.9.1, there exists $n' > N$ such that $\rho(x_{n'}, c) < \frac{1}{2}\varepsilon$. Therefore, for any $n > N$,

$$\rho(x_n, c) \leqslant \rho(x_n, x_{n'}) + \rho(x_{n'}, c) < \varepsilon.$$

Thus (x_n) converges to c.

THEOREM 4.1.4. *Every fundamental sequence of a metric space is bounded.*

Proof. Let (x_n) be a fundamental sequence of (X, ρ). Then there exists N such that $\rho(x_m, x_n) < 1$ for all $m, n \geqslant N$, so $x_n \in S(x_N, 1)$ for all $n \geqslant N$. There remain $x_1, x_2, \ldots, x_{N-1}$ which may not be in this sphere; we therefore define another sphere, centre x_N, but with suitably larger radius. Thus, let

$$r > \max\{1, \rho(x_1, x_N), \rho(x_2, x_N), \ldots, \rho(x_{N-1}, x_N)\},$$

so $x_n \in S(x_N, r)$ for all n. Hence the sequence is bounded.

It is clearly a corollary of this result that every convergent sequence of a metric space is bounded.

EXERCISES 4.1

1. Let (a_n), (b_n) be sequences in (X, ρ) such that (a_n) is a fundamental sequence and $\rho(a_n, b_n) \to 0$ as $n \to \infty$. Show that

   (i) (b_n) is a fundamental sequence in (X, ρ);

   (ii) (b_n) converges to a limit l if and only if (a_n) also converges to l.

2. Show that any metric space with only a finite number of points is complete.

3. Show that the metric space (N, ρ) defined in Exercise 3.4.6 is not complete.

4. Show that $((0, 1), \mathsf{d})$ is not complete. Deduce that if (X, ρ), (X', ρ') are an arbitrary pair of homeomorphic spaces, then completeness of one space does *not* imply completeness of the other. [Note, however, the result of the next exercise.]

5. Let (X, ρ), (X', ρ') be two metric spaces and $f:(X, \rho) \to (X', \rho')$ be uniformly continuous. If (x_n) is a fundamental sequence in (X, ρ), show that $(f(x_n))$ is a fundamental sequence in (X', ρ'). If, furthermore, f is a bijective function such that f^{-1} is also uniformly continuous, prove that (X, ρ) is complete if and only if (X', ρ') is complete.

6. Define $\rho: \mathsf{R} \times \mathsf{R} \to \mathsf{R}$ by

$$\rho(x, y) = \frac{|x-y|}{\sqrt{(1+x^2)}\sqrt{(1+y^2)}}.$$

Show that (R, ρ) is a metric space which is not complete. Show also that the identity mapping $i:(\mathsf{R}, \mathsf{d}) \to (\mathsf{R}, \rho)$ is uniformly continuous. [Thus a uniformly continuous mapping does not necessarily map a complete space into a complete space.]

4.2 Examples concerning the completeness of metric spaces

The metric spaces of §2.2 will now be examined to see whether or not they are complete. The reader should refer back to §2.2 for the details of definition of each space; he is also reminded of the convention introduced at the beginning of §2.4.

(i), (ii) $X = \mathsf{R}, \mathsf{R}^m$ respectively, with ρ the corresponding Euclidean metric; then these metric spaces are complete—by the Cauchy principle of convergence in 1, m dimensions respectively.

(iii) $X = \mathsf{R}^m$, and

$$\rho(x, y) = \left\{\sum_{i=1}^{m} |x_i - y_i|^p\right\}^{1/p}.$$

Let $(x^{(n)})$ be a fundamental sequence where $x^{(n)} = (x_1^{(n)}, \ldots, x_m^{(n)})$. Then given any $\varepsilon > 0$ there exists N such that

$$\sum_{i=1}^{m} |x_i^{(n)} - x_i^{(n')}|^p < \varepsilon^p, \quad \text{for all } n, n' > N. \tag{4.2.1}$$

Hence for each i, $|x_i^{(n)} - x_i^{(n')}| < \varepsilon$ for all $n, n' > N$; therefore, by the Cauchy principle of convergence there exists a real number, x_i say, such that $x_i^{(n)} \to x_i$ as $n \to \infty$. Let $x = (x_1, \ldots, x_m)$; it will be shown that $x^{(n)} \to x$ in (X, ρ), so the latter is complete. In doing this we take particular care with the details in order to throw further light on the similar, but more difficult, argument which is used below in (vi).

Let n be any integer greater than N and set

$$y_{n'} = \sum_{i=1}^{m} |x_i^{(n)} - x_i^{(n')}|^p;$$

therefore, from (4.2.1), $y_{n'} < \varepsilon^p$ for all $n' > N$, and hence

$$\lim_{n' \to \infty} y_{n'} \leqslant \varepsilon^p.$$

Since

$$\lim_{n' \to \infty} \sum_{i=1}^{m} |x_i^{(n)} - x_i^{(n')}|^p = \sum_{i=1}^{m} |x_i^{(n)} - x_i|^p, \tag{4.2.2}$$

it follows that

$$\sum_{i=1}^{m} |x_i^{(n)} - x_i|^p \leqslant \varepsilon^p \quad \text{for all } n \geqslant N;$$

thus $x^{(n)} \to x$ as $n \to \infty$ in (X, ρ).

(iv) $X = \mathsf{R}^m$, and $\rho(x, y) = \max_i |x_i - y_i|$. Completeness is established exactly as in (iii).

(v) X is any non-empty set, and ρ is the standard discrete metric. Let (x_n) be a fundamental sequence; then there exists N such that $\rho(x_m, x_n) < 1$ for all $m, n \geqslant N$, so $x_n = x_N$ for all $n \geqslant N$. Thus any fundamental sequence must be of the form

$$(x_1, x_2, \ldots, x_N, x_N, x_N, \ldots)$$

which clearly converges to x_N (in X). Thus (X, ρ) is complete.

However *not* every discrete metric space is complete; see, for example, Exercise 4.1.3.

(vi) $X = \ell^p$, and

$$\rho(x, y) = \left\{ \sum_{i=1}^{\infty} |x_i - y_i|^p \right\}^{1/p}.$$

Then (ℓ^p, ρ) is complete; the proof of this is similar to, but not as easy as, that of (iii).

Let $(x^{(n)})$ be a fundamental sequence where $x^{(n)} = (x_1^{(n)}, x_2^{(n)}, \ldots)$. Then given any $\varepsilon > 0$ there exists N such that

$$\sum_{i=1}^{\infty} |x_i^{(n)} - x_i^{(n')}|^p < \varepsilon^p, \quad \text{for all } n, n' > N. \tag{4.2.3}$$

Thus for any positive integer m,

$$\sum_{i=1}^{m} |x_i^{(n)} - x_i^{(n')}|^p < \varepsilon^p, \quad \text{for all } n, n' > N. \tag{4.2.4}$$

Also for each i, $|x_i^{(n)} - x_i^{(n')}| < \varepsilon$, for all $n, n' > N$; as before it follows that $x_i^{(n)}$ must tend to a real limit, x_i say, as $n \to \infty$. Let $x = (x_1, x_2, \ldots)$; it will be shown that $x_n \to x$ in (X, ρ). We cannot use exactly the same argument as in (iii) since (4.2.2) no longer follows immediately when m is replaced by ∞; why? We modify the argument as follows.

From (4.2.4), letting $n' \to \infty$, it follows that

$$\sum_{i=1}^{m} |x_i^{(n)} - x_i|^p \leqslant \varepsilon^p, \tag{4.2.5}$$

for any positive integer m. Now let the left side of (4.2.5) be denoted by s_m; then (s_m) is a positive bounded increasing sequence, so has a finite limit, s say, as $m \to \infty$ and $s \leqslant \varepsilon^p$. Hence

$$\left\{\sum_{i=1}^{\infty} |x_i^{(n)} - x_i|^p\right\}^{1/p} \leqslant \varepsilon, \quad \text{for all } n > N;$$

thus $x^{(n)} \to x$ as $n \to \infty$.

Finally we must show that $x \in \ell^p$; writing

$$x_i = (x_i - x_i^{(n)}) + x_i^{(n)},$$

then by Minkowski's inequality

$$\{\textstyle\sum_i |x_i|^p\}^{1/p} \leqslant \{\sum_i |x_i - x_i^{(n)}|^p\}^{1/p} + \{\sum_i |x_i^{(n)}|^p\}^{1/p} < \infty,$$

so $x \in \ell^p$.

(vii) $X = m$ and $\rho(x, y) = \sup_i |x_i - y_i|$. An argument similar to, but rather simpler than, that used in (vi) shows that (m, ρ) is complete.

(viii) $X = \mathscr{C}(I)$, and $\rho(x, y) = \sup_t |x(t) - y(t)|$. Let (x_n) be a fundamental sequence; then given any $\varepsilon > 0$ there exists N such that

$$\sup_t |x_n(t) - x_m(t)| < \varepsilon, \quad \text{for all } m, n > N.$$

Hence for each t in I, $|x_n(t) - x_m(t)| < \varepsilon$ for all $m, n > N$, where N is independent of t. Therefore by the Cauchy principle of uniform convergence (see Theorem 1.6.9), it follows that there exists a function x such that $x_n \to x$ uniformly over I; furthermore (by Theorem 1.6.7) $x \in \mathscr{C}(I)$. This establishes completeness.

(ix) $X = \mathscr{B}(I)$, and ρ is again the supremum metric. Then $(\mathscr{B}(I), \rho)$ is complete.

(x) $X = \mathscr{C}(I)$, and

$$\rho(x, y) = \int_a^b |x(t) - y(t)| \, dt; \tag{4.2.6}$$

then $(\mathscr{C}(I), \rho)$ is not complete.

For, let $a < c < b$, and n be a sufficiently large positive integer such that $a < c - 1/n$; then define $x_n : I \to \mathrm{R}$ by

$$x_n(t) = \begin{cases} 0 & \text{if } a \leqslant t \leqslant c - 1/n, \\ nt - nc + 1 & \text{if } c - 1/n < t \leqslant c, \\ 1 & \text{if } c < t \leqslant b. \end{cases}$$

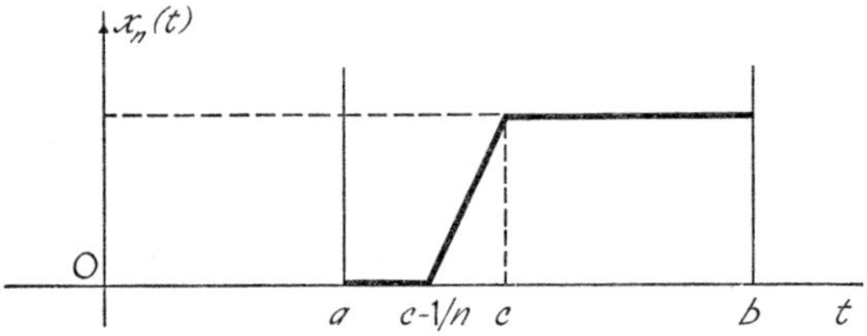

Clearly x_n is continuous over I, and

$$\rho(x_n, x_m) \leqslant \int_{c-1/n}^{c} x_n(t) \, dt + \int_{c-1/m}^{c} x_m(t) \, dt = \tfrac{1}{2}(n^{-1} + m^{-1}),$$

so (x_n) is a fundamental sequence.

Suppose now that there exists a continuous function x such that $\rho(x_n, x) \to 0$; it will be shown that this leads to a contradiction. Since

$$\rho(x_n, x) = \int_a^{c-1/n} |x(t)| \, dt + \int_{c-1/n}^{c} |x_n(t) - x(t)| \, dt + \int_c^b |1 - x(t)| \, dt,$$

and $\rho(x, x_n) \to 0$, it follows that

$$x(t) = \begin{cases} 0 & \text{for } t \text{ in } [a, c) \\ 1 & \text{for } t \text{ in } (c, b]. \end{cases}$$

Clearly x is not continuous, so we have a contradiction.

Consequently the metric space is not complete.

It is therefore seen that if we associate the supremum metric with $\mathscr{C}(I)$ we obtain a complete space; on the other hand if we associate the integral metric (4.2.6) with $\mathscr{C}(I)$ we obtain a metric space which is not complete. For this reason, as far as we are concerned in the present text, the supremum metric, or a slight variation of it which we introduce later, is the one which we shall find most useful to associate with $\mathscr{C}(I)$.

However in a more general situation where the integral (4.2.6) is taken in the Lebesgue sense over the set of all Lebesgue integrable functions the resulting metric space is complete.

EXERCISES 4.2

1. Using the result of Exercise 4.1.5 and the example of § 3.4 show that the metric spaces defined in §2.2 (iii), (iv) are complete.

2. Using Exercise 2.4.1 show that the metric spaces (X, ρ), (X, ρ') defined in Exercise 2.2.2 are complete if and only if the spaces (X_i, ρ_i), $i = 1, \ldots, n$ are all complete.

3. Let (m, ρ) be the metric space defined in §2.2 (vii); let $c \subset m$ be the subset consisting of all convergent sequences. By Exercise 2.6.6 it follows that c is closed in (m, ρ), so (c, ρ_c) is complete.

Let c_0 be the subset consisting of all convergent sequences whose limit is zero; show that (c_0, ρ_{c_0}) is complete.

Let n be the subset consisting of all sequences (x_i) such that only a finite number of the x_i are non-zero; show that (n, ρ_n) is not complete.

4. Using Exercise 2.4.2 show that the metric spaces defined in Exercises 2.2.3, 2.2.4 are complete.

5. Show that the metric spaces defined in Exercises 2.1.5, 2.1.6 are complete.

6. Show that the metric space defined in Exercise 2.4.4 is complete.

4.3 Cantor's intersection theorem

First the reader is reminded of the following result which holds in R^m (with the Euclidean metric).

PROPOSITION 4.3.1 (Cantor's intersection theorem). *Let* $(\bar{S}_n)$ *be a nested*† *sequence of closed spheres in* R^m; *let* r_n *be the radius of* $\bar{S}_n$. *If*

† A sequence (X_n) of sets is said to be *nested* if $X_n \supseteq X_{n+1}$ for each n.

$r_n \to 0$ as $n \to \infty$, then there exists a unique point x common to all of the spheres, that is

$$\bigcap_{n=1}^{\infty} \bar{S}_n = \{x\}.$$

The proof of this result depends essentially on the completeness of R^m. The result does not hold in the (non-complete) metric space (Q, d). For let x be any irrational in R; by the result quoted in the footnote on p. 47, there exists an increasing sequence (s_n) of rationals and a decreasing sequence (t_n) of rationals such that both sequences converge to x. Now set

$$\bar{S}_n = \{q : s_n \leqslant q \leqslant t_n, q \in \mathsf{Q}\}.$$

Then clearly $(\bar{S}_n)$ is a nested sequence of closed spheres in (Q, d). Furthermore there is no point common to all the $\bar{S}_n$; why?

Cantor's theorem can be generalized to any complete metric space. This we now do; a two-way result is established.

THEOREM 4.3.1. *Let (X, ρ) be a metric space. Then the following statements are equivalent:*

(i) *(X, ρ) is complete;*

(ii) *every nested sequence $(\bar{S}_n)$ of closed spheres (in (X, ρ)), with radii tending to zero, has exactly one point in the intersection*

$$\bigcap_{n=1}^{\infty} \bar{S}_n.$$

Proof. Assume (i); let $(\bar{S}_n)$ be a nested sequence of closed spheres, and let $\bar{S}_n = \bar{S}(x_n, r_n)$ where $r_n \to 0$ as $n \to \infty$.

If $m > n$, then $\bar{S}_m \subseteq \bar{S}_n$, so $x_m \in \bar{S}_n$, and thus $\rho(x_m, x_n) \leqslant r_n$; hence $\rho(x_m, x_n) \to 0$ as $m, n \to \infty$, that is (x_n) is a fundamental sequence. Since (X, ρ) is complete, the sequence (x_n) converges to a limit, x say, in X. Since all the points of the sequence $(x_n, x_{n+1}, x_{n+2}, \ldots)$ are in $\bar{S}_n$, and since $\bar{S}_n$ is closed, it follows that $x \in \bar{S}_n$. This is true for every n, so

$$x \in \bigcap_{n=1}^{\infty} \bar{S}_n.$$

Next is shown that x is the only point common to all the spheres. For suppose $y \in \bar{S}_n$ for all n; then

$$\rho(x, y) \leqslant \rho(x, x_n) + \rho(x_n, y) \leqslant 2r_n$$

for all n. Since $r_n \to 0$, it follows that $\rho(x, y) = 0$, so $x = y$.

Now assume (ii), and let (x_n) be a fundamental sequence in (X, ρ). Then, given any $\varepsilon > 0$, there exists N such that $\rho(x_m, x_n) < \varepsilon$ for all $m, n \geqslant N$ and, in particular, $\rho(x_N, x_m) < \varepsilon$ for all $m \geqslant N$.

Let n_1 be defined so that $\rho(x_{n_1}, x_m) < \frac{1}{2}$ for all $m \geqslant n_1$; denote $\bar{S}_1 = \bar{S}(x_{n_1}, 1)$. Let n_2 $(> n_1)$ be defined such that $\rho(x_{n_2}, x_m) < \frac{1}{4}$ for all $m \geqslant n_2$; denote $\bar{S}_2 = \bar{S}(x_{n_2}, \frac{1}{2})$. Then if $x \in \bar{S}_2$

$$\rho(x, x_{n_1}) \leqslant \rho(x, x_{n_2}) + \rho(x_{n_2}, x_{n_1}) < 1,$$

so that $x \in \bar{S}_1$; hence $\bar{S}_1 \supseteq \bar{S}_2$. Next let $n_3 (> n_2)$ be defined such that $\rho(x_{n_3}, x_m) < \frac{1}{8}$ for all $m \geqslant n_3$; denote $\bar{S}_3 = \bar{S}(x_{n_3}, \frac{1}{4})$. It is easily seen, as above, that $\bar{S}_2 \supseteq \bar{S}_3$.

Proceeding in this way (that is, by induction) we can define a nested sequence of closed spheres $(\bar{S}_k)$ where the radius of $\bar{S}_k$ is $2^{-(k-1)}$. Then by hypothesis there exists a unique point x such that

$$\bigcap_{k=1}^{\infty} \bar{S}_k = \{x\}.$$

Since $\rho(x_{n_k}, x) \leqslant 2^{-(k-1)}$ it follows that $x_{n_k} \to x$ as $k \to \infty$; by Theorem 4.1.3 it follows that $x_n \to x$ as $n \to \infty$.

Thus any fundamental sequence of (X, ρ) is convergent.

EXERCISES 4.3

1. Show that Theorem 4.3.1 can be reformulated as follows.

Let (X, ρ) be a metric space. Then the following statements are equivalent:

(i) (X, ρ) is complete;

(ii) every nested sequence (A_n) of non-empty closed sets whose diameters (see Exercise 3.5.4) tend to zero, has exactly one point in the intersection

$$\bigcap_{n=1}^{\infty} A_n.$$

Taking $A_n = (0, n^{-1})$ and $A_n = [n, \infty)$ respectively, show that neither of the conditions: all the sets A_n are closed, $d(A_n) \to 0$, may be omitted from (ii).

2. Define $\rho: \mathsf{N} \times \mathsf{N} \to \mathsf{R}$ by

$$\rho(m, n) = \begin{cases} 1 + (m+n)^{-1} & \text{if } m \neq n \\ 0 & \text{if } m = n; \end{cases}$$

verify that (N, ρ) is a complete metric space. Show that $(\bar{S}(n, 1+(2n)^{-1}))$ is a nested sequence of closed spheres whose radii tend to a non-zero limit, but the intersection of the spheres is empty.

[Thus in Theorem 4.3.1 it is not possible to omit the condition that the radii of the spheres tend to zero. See, however, Exercise 5.2.5.]

4.4 The completion of incomplete metric spaces

It was explained in §4.1 that (Q, d) is not complete but (R, d) is complete; thus (Q, d) is a subspace of a complete metric space. In the imprecise, but suggestive, language of §4.1 we say that the gaps of (Q, d) may be filled by regarding Q as a subset of a larger space, namely (R, d). A question which arises from this is 'For any metric space (X, ρ) which is not complete, does there always exist a complete metric space of which (X, ρ) is a subspace?'. In brief the answer to this is 'to all intents and purposes, yes'. In the process of making this precise and establishing the assertion, we go further and show how to construct a complete metric space, which we shall denote by $(\tilde{X}, \tilde{\rho})$; (X, ρ) is not truly a subspace but is, as far as we are concerned (in relation to all its metric space properties), equivalent to a subspace of $(\tilde{X}, \tilde{\rho})$. We describe this situation by saying that (X, ρ) is embedded in $(\tilde{X}, \tilde{\rho})$.

In order to carry out this procedure we look at the method by which the set of rationals is embedded in the set of reals. There are several ways in which this can be done. Firstly there is Dedekind's method of cuts; however this depends essentially on the order relation (the relation $<$) of the rationals, and so the method cannot be generalized to an arbitrary metric space (since this does not necessarily have an order relation). Secondly there is Cantor's method in which the reals are constructed from the fundamental sequences of rationals (see Cohen and Ehrlich (1963), Chapter 4). This method does not involve any ideas which are not capable of generalization to any metric space; the method which we describe will be based on the same procedure.

First it is necessary to define some elementary terms.

DEFINITION 4.4.1. Let (X, ρ), (X', ρ') be two metric spaces, and $f: X \to X'$ be a bijection. Then f is said to be an *isometry* (or an *isometric mapping of* (X, ρ) *to* (X', ρ')) if, for all x, y in X.

$$\rho'(f(x), f(y)) = \rho(x, y).$$

More briefly we shall then say that $f:(X, \rho) \to (X', \rho')$ is an isometry. An isometry is thus a bijection between two metric spaces which preserves distances. Clearly $f^{-1}:(X', \rho') \to (X, \rho)$ is an isometry if $f:(X, \rho) \to (X', \rho')$ is an isometry.

Two metric spaces (X, ρ), (X', ρ') are said to be *isometric* if there exists an isometry of (X, ρ) to (X', ρ'). Any result which has been proved for a metric space (X, ρ) and which depends only on the distances between elements of X, for example, convergence, completeness, etc., will give a corresponding result concerning any isometric space (X', ρ'). Thus for such results isometric spaces may be considered as equivalent.

It is clear that

(i) any isometric mapping is continuous, and indeed is a homeomorphism;

(ii) isometry is an equivalence relation.

Next we say precisely what we mean by embedding.

DEFINITION 4.4.2. If (X', ρ') is a metric space, X'' is a subset of X', and (X, ρ) is a metric space such that (X, ρ), (X'', ρ') are isometric, then (X, ρ) is said to be *isometrically embedded* in (X', ρ').

The diagram illustrates this definition.

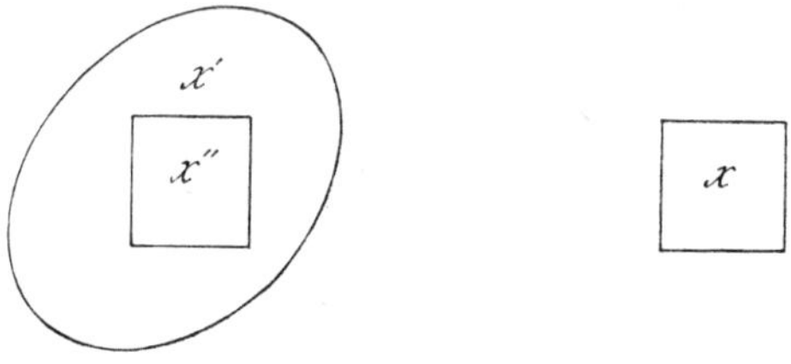

Finally one other term is required.

DEFINITION 4.4.3. In a metric space (X, ρ) any two fundamental sequences (x_n), (y_n) such that $\rho(x_n, y_n) \to 0$ as $n \to \infty$ are said to be *equivalent*; this is written as $(x_n) \sim (y_n)$.

This property is indeed an equivalence relation; verification is trivial. Therefore (by Theorem 1.3.1) the set of all fundamental sequences may be partitioned into equivalence classes in such a way that all the sequences which are mutually equivalent belong to the same class.

The main result is now given; the proof is long, so it has been divided into stages. The theorem essentially amounts to the statement that any metric space (X, ρ) can be isometrically embedded in a complete metric space $(\tilde{X}, \tilde{\rho})$. (In this section we shall write $(\tilde{Y}, \tilde{\rho})$ in place of $(\tilde{Y}, \tilde{\rho}_{\tilde{Y}})$, where $\tilde{Y} \subset \tilde{X}$.)

THEOREM 4.4.1. *Let (X, ρ) be any metric space; then there exists a complete metric space $(\tilde{X}, \tilde{\rho})$ which contains a subspace $(\tilde{Y}, \tilde{\rho})$ such that*

(i) *(X, ρ) is isometric to $(\tilde{Y}, \tilde{\rho})$,*
(ii) *$\tilde{Y}$ is everywhere dense in $(\tilde{X}, \tilde{\rho})$.*

Proof. I. *To construct the metric space $(\tilde{X}, \tilde{\rho})$.*

Let $\tilde{X}$ be the collection of all equivalence classes of equivalent (fundamental) sequences of X. Let $\tilde{x}, \tilde{y}$ be two elements of $\tilde{X}$; from these classes choose, respectively, two arbitrary sequences (x_n), (y_n) so $(x_n) \in \tilde{x}$, $(y_n) \in \tilde{y}$. It will be shown that $(\rho(x_n, y_n))$ is a fundamental sequence of reals.

By the inequality (2.3.1) we have

$$|\rho(x_m, y_m) - \rho(x_n, y_n)| \leqslant \rho(x_m, x_n) + \rho(y_m, y_n);$$

since (x_n), (y_n) are fundamental, given any $\varepsilon > 0$ there exists N such that $\rho(x_m, x_n) < \frac{1}{2}\varepsilon$ and $\rho(y_m, y_n) < \frac{1}{2}\varepsilon$ for all $m, n > N$, so

$$|\rho(x_m, y_m) - \rho(x_n, y_n)| < \varepsilon$$

for all $m, n > N$. Thus the sequence $(\rho(x_n, y_n))$ of reals is fundamental and so (by the Cauchy principle) possesses a limit in R; denote this limit by $\tilde{\rho}(\tilde{x}, \tilde{y})$, that is set

$$\tilde{\rho}(\tilde{x}, \tilde{y}) = \lim_{n \to \infty} \rho(x_n, y_n). \tag{4.4.1}$$

It is easy to see that $\tilde{\rho}(\tilde{x}, \tilde{y})$ is independent of the sequences (x_n), (y_n) which are selected from $\tilde{x}, \tilde{y}$. For suppose $(x_n), (x'_n) \in \tilde{x}$, $(y_n), (y'_n) \in \tilde{y}$; then

$$|\rho(x'_n, y'_n) - \rho(x_n, y_n)| \leqslant \rho(x_n, x'_n) + \rho(y_n, y'_n)$$

(again by (2.3.1)), so on letting $n \to \infty$ it follows that

$$\lim_{n \to \infty} \rho(x'_n, y'_n) = \lim_{n \to \infty} \rho(x_n, y_n),$$

which establishes our assertion.

Next it is shown that $(\tilde{X}, \tilde{\rho})$ is a metric space.

Firstly it is clear that $\tilde{\rho}(\tilde{x}, \tilde{y}) \geqslant 0$; also

$$\tilde{\rho}(\tilde{x}, \tilde{x}) = \lim_{n\to\infty} \rho(x_n, x_n) = 0.$$

Suppose $\tilde{\rho}(\tilde{x}, \tilde{y}) = 0$, so $\rho(x_n, y_n) \to 0$ as $n \to \infty$; hence $(x_n) \sim (y_n)$, and so $\tilde{x} = \tilde{y}$. Thus $\tilde{\rho}(\tilde{x}, \tilde{y}) \geqslant 0$, with equality if and only if $\tilde{x} = \tilde{y}$.

Secondly, and trivially, $\tilde{\rho}(\tilde{x}, \tilde{y}) = \tilde{\rho}(\tilde{y}, \tilde{x})$ for all $\tilde{x}, \tilde{y}$ in X.

Lastly, for any $\tilde{x}, \tilde{y}, \tilde{z}$ in X, if $(x_n) \in \tilde{x}$, $(y_n) \in \tilde{y}$, $(z_n) \in \tilde{z}$, since

$$\rho(x_n, y_n) \leqslant \rho(x_n, z_n) + \rho(z_n, y_n),$$

it follows that

$$\tilde{\rho}(\tilde{x}, \tilde{y}) \leqslant \tilde{\rho}(\tilde{x}, \tilde{z}) + \tilde{\rho}(\tilde{z}, \tilde{y}).$$

Hence $(\tilde{X}, \tilde{\rho})$ is a metric space.

II. *To show that* (X, ρ) *is isometrically embedded in* $(\tilde{X}, \tilde{\rho})$.

Let $\tilde{Y}$ denote the subset of $\tilde{X}$ such that $\tilde{x} \in \tilde{Y}$ if and only if each fundamental sequence of $\tilde{x}$ is convergent. It is easily verified that

(i) if $(x_n) \in \tilde{x}$ converges to some element x in X, then every fundamental sequence of $\tilde{x}$ also converges to x;

(ii) furthermore, the fundamental sequence $(x, x, \ldots, x, \ldots)$ (known as a *constant* sequence) will then also belong to $\tilde{x}$;

(iii) this is the only constant sequence in $\tilde{x}$.

These facts enable us to exhibit a natural bijection $f: X \to \tilde{Y}$; it is defined as follows. Let $x \in X$, and let $\tilde{x}$ be that element of $\tilde{X}$ which contains the constant sequence $(x, x, \ldots)$; then $\tilde{x} \in \tilde{Y}$. Let $f(x) = \tilde{x}$ for all x in X. By (ii) and (iii) this is a bijection of X onto $\tilde{Y}$. Furthermore, if $x, y \in X$, then $\tilde{\rho}(\tilde{x}, \tilde{y}) = \rho(x, y)$, so (X, ρ), $(\tilde{Y}, \tilde{\rho})$ are isometric.

III. *To show that* $\tilde{Y}$ *is everywhere dense in* $(\tilde{X}, \tilde{\rho})$.

Let $\tilde{x} \in \tilde{X}$ and let (x_n) be a fundamental sequence in $\tilde{x}$; then given any $\varepsilon > 0$ there exists N such that $\rho(x_m, x_N) < \varepsilon$ for all $m \geqslant N$. Denote the equivalence class to which the constant sequence $(x_N, x_N, \ldots)$ belongs by $\tilde{x}_N$, so that $\tilde{x}_N \in \tilde{Y}$. Now

$$\tilde{\rho}(\tilde{x}, \tilde{x}_N) = \lim_{p\to\infty} \rho(x_p, x_N) \leqslant \varepsilon.$$

Hence, by Lemma 2.10.1, $\tilde{Y}$ is everywhere dense.

IV. *To show that* $(\tilde{X}, \tilde{\rho})$ *is complete.*

This is more difficult to establish. In accordance with the remark at the beginning of §2.4, let $(\tilde{x}^{(n)})$ be an arbitrary fundamental sequence in $(\tilde{X}, \tilde{\rho})$; our aim is to show that this sequence has a limit $\tilde{x}$ in $\tilde{X}$.

Since $\tilde{Y}$ is everywhere dense in $(\tilde{X}, \tilde{\rho})$, for each integer n there exists an element $\tilde{y}^{(n)}$ of $\tilde{Y}$ such that $\tilde{\rho}(\tilde{x}^{(n)}, \tilde{y}^{(n)}) < n^{-1}$. Since $(\tilde{x}^{(n)})$ is a fundamental sequence, given any $\varepsilon > 0$ there exists N such that $\tilde{\rho}(\tilde{x}^{(m)}, \tilde{x}^{(n)}) < \frac{1}{2}\varepsilon$ for all $m, n > N$. Then it follows that $(\tilde{y}^{(n)})$ is also a fundamental sequence; for

$$\begin{aligned}\tilde{\rho}(\tilde{y}^{(m)}, \tilde{y}^{(n)}) &\leqslant \tilde{\rho}(\tilde{y}^{(m)}, \tilde{x}^{(m)}) + \tilde{\rho}(\tilde{x}^{(m)}, \tilde{x}^{(n)}) + \tilde{\rho}(\tilde{x}^{(n)}, \tilde{y}^{(n)}) \\ &< m^{-1} + \tfrac{1}{2}\varepsilon + n^{-1},\end{aligned}$$

for all $m, n > N'$, where $N' = \max(N, 4/\varepsilon)$.

Since $\tilde{y}^{(n)} \in \tilde{Y}$, it follows that all the (fundamental) sequences in $\tilde{y}^{(n)}$ are convergent, having a common limit, say y_n; in particular the constant sequence $(y_n, y_n, \ldots)$ belongs to $\tilde{y}^{(n)}$. Now, by (4.4.1),

$$\rho(y_m, y_n) = \tilde{\rho}(\tilde{y}^{(m)}, \tilde{y}^{(n)}) < \varepsilon \qquad (4.4.2)$$

for all $m, n > N'$, so (y_n) is a fundamental sequence of (X, ρ). Hence it belongs to some element of $\tilde{X}$; call this element $\tilde{x}$.

Lastly it will be shown that the fundamental sequence $(\tilde{x}^{(n)})$ converges to $\tilde{x}$. Now

$$\begin{aligned}\tilde{\rho}(\tilde{x}^{(n)}, \tilde{x}) &\leqslant \tilde{\rho}(\tilde{x}^{(n)}, \tilde{y}^{(n)}) + \tilde{\rho}(\tilde{y}^{(n)}, \tilde{x}) \\ &< n^{-1} + \tilde{\rho}(\tilde{y}^{(n)}, \tilde{x});\end{aligned}$$

but by (4.4.1) and (4.4.2)

$$\tilde{\rho}(\tilde{y}^{(n)}, \tilde{x}) = \lim_{p \to \infty} \rho(y_n, y_p) \leqslant \varepsilon$$

for all $n > N'$. Hence $\tilde{\rho}(\tilde{x}^{(n)}, \tilde{x}) < 2\varepsilon$ for all n sufficiently large and so $(\tilde{x}^{(n)})$ converges to $\tilde{x}$ in $(\tilde{X}, \tilde{\rho})$.

This concludes the proof.

We make some remarks.

(i) The above proof depends essentially on the completeness of R (with Euclidean metric); thus the proof does not provide a method for constructing the real number system from the rationals, although it was motivated by Cantor's construction of the reals.

(ii) If, in the notation of Theorem 4.4.1, (X, ρ) is complete, then (X, ρ) and $(\tilde{X}, \tilde{\rho})$ are isometric and can be identified.

(iii) There exist other methods of obtaining a completion $(\tilde{X}, \tilde{\rho})$ of (X, ρ); one such method is outlined in Exercise 4.4.4. In this, (X, ρ) will again be found to be isometric to an everywhere dense subset of $(\tilde{X}, \tilde{\rho})$. By Theorem 4.5.1 the completions obtained in the the two ways will be isometric, so may be identified.

EXERCISES 4.4

1. Show that two convergent sequences of (X, ρ) are equivalent if and only if they have the same limit.

2. If the metric spaces (X, ρ), (X', ρ') are isometric, show that (X, ρ) is complete if and only if (X', ρ') is complete.

3. Show that the sequences (a_n), (b_n) are fundamental and equivalent if and only if they are both subsequences of some fundamental sequence (c_n).

4. Let (X, ρ) be a metric space and let $x_0 \in X$. Let X^* denote the set of all functions from (X, ρ) to (R, d) which are bounded and continuous over X.

(i) For each x in X define a function f_x by

$$f_x(y) = \rho(y, x) - \rho(y, x_0).$$

Show that

$$|f_x(y)| \leqslant \rho(x, x_0), \quad |f_x(y) - f_x(y')| \leqslant 2\rho(y, y');$$

deduce that $f_x \in X^*$.

(ii) Define $F: X \to X^*$ by $F(x) = f_x$. Associate with X^* the supremum metric

$$\tilde{\rho}(g, h) = \sup_{x \in X} |g(x) - h(x)|.$$

Verify that $(X^*, \tilde{\rho})$ is a complete metric space.

Define $\tilde{X}$ to be the closure of $F(X)$ in $(X^*, \tilde{\rho})$, so $(\tilde{X}, \tilde{\rho})$ is complete and $F(X)$ is everywhere dense in $(\tilde{X}, \tilde{\rho})$.

(iii) Show that F is injective, and that

$$|f_x(y)-f_{x'}(y)| \leqslant \rho(x, x')$$

for all x, x', y in X, and that equality occurs. Deduce that

$$\tilde{\rho}(F(x), F(x')) = \rho(x, x'),$$

and so (X, ρ), $(F(X), \tilde{\rho})$ are isometric.

4.5 The uniqueness of $(\tilde{X}, \tilde{\rho})$ up to isometry

The next question to be discussed is whether a metric space is embedded uniquely, in any sense, in a complete metric space.

Since (Q, d) can be embedded in the complete metric space $(\mathsf{R}^m, \mathsf{d})$, for $m = 1, 2, \ldots$, where d is the relevant Euclidean metric in each case, it follows that the embedding of a metric space in a complete metric space is not unique. However it is clear that by considering (Q, d) as being embedded in $(\mathsf{R}^m, \mathsf{d})$, $m \geqslant 2$ we are embedding (Q, d) in unnecessarily large metric spaces. Therefore we restrict (in an obvious way) the set of completions of a given metric space, and ask again the question concerning uniqueness for these completions. The answer, now in the affirmative, is contained in the next result. (In this section we write $(\tilde{Y}, \tilde{\rho})$ in place of $(\tilde{Y}, \tilde{\rho}_{\tilde{Y}})$ and $(\tilde{Y}', \tilde{\rho}')$ in place of $(\tilde{Y}', \tilde{\rho}_{Y'})$.)

THEOREM 4.5.1. *Let (X, ρ) be a metric space; let $(\tilde{X}, \tilde{\rho})$ be the completion of it constructed in the proof of Theorem* 4.4.1, *and $\tilde{Y}$ be the subset of $\tilde{X}$ defined in* §4.4, *so (X, ρ) and $(\tilde{Y}, \tilde{\rho})$ are isometric.*

Suppose $(\tilde{X}', \tilde{\rho}')$ is any other complete metric space such that

(i) *(X, ρ) is isometrically embedded in $(\tilde{X}', \tilde{\rho}')$, and let $\tilde{Y}'$ be that subset of $\tilde{X}'$ such that (X, ρ) and $(\tilde{Y}', \tilde{\rho}')$ are isometric, and*

(ii) *$\tilde{Y}'$ is everywhere dense in $(\tilde{X}', \tilde{\rho}')$;*

then $(\tilde{X}, \tilde{\rho})$ and $(\tilde{X}', \tilde{\rho}')$ are isometric.

Proof. By Theorem 4.4.1, (X, ρ) is isometric to $(\tilde{Y}, \tilde{\rho})$, and by the hypotheses of the present theorem, (X, ρ) is isometric to $(\tilde{Y}', \tilde{\rho}')$; therefore $(\tilde{Y}, \tilde{\rho})$ is isometric to $(\tilde{Y}', \tilde{\rho}')$ since isometry is an equivalence relation. Let $g: \tilde{Y} \to \tilde{Y}'$ be an isometric mapping of $(\tilde{Y}, \tilde{\rho})$ to $(\tilde{Y}', \tilde{\rho}')$.

An isometry of $(\tilde{X}, \tilde{\rho})$ to $(\tilde{X}', \tilde{\rho}')$ is obtained by extending the domain of g to $\tilde{X}$ in the way which arises naturally from the fact that

$\tilde{Y}$ is everywhere dense in $(\tilde{X}, \tilde{\rho})$. It is then shown that the extended mapping so obtained is also an isometry. The details follow.

In the remainder of the proof it is convenient to denote a sequence in $\tilde{X}$ by $(\tilde{x}_n)$ rather than $(\tilde{x}^{(n)})$ as was done in §4.4; no confusion should arise.

Let $\tilde{x} \in \tilde{X}$, so there exists a sequence $(\tilde{x}_n)$ in $\tilde{Y}$ which converges to $\tilde{x}$ in $(\tilde{X}, \tilde{\rho})$. Then $(\tilde{x}_n)$ is a fundamental sequence of $(\tilde{Y}, \tilde{\rho})$; hence $(g(\tilde{x}_n))$ is a fundamental sequence of $(\tilde{Y}', \tilde{\rho}')$, and so it has a limit, say $\tilde{x}'$, in $(\tilde{X}', \tilde{\rho}')$.

Extend the mapping $g: \tilde{Y} \to \tilde{Y}'$ to a mapping $G: \tilde{X} \to \tilde{X}'$ by defining $G(\tilde{x}) = \tilde{x}'$. In order that this definition be meaningful and agree with g on $\tilde{Y}$ it is necessary to verify that it is independent of the choice of sequence $(\tilde{x}_n)$. To do this, suppose that $(\tilde{y}_n)$ is another sequence in $\tilde{Y}$ which also converges to $\tilde{x}$; then, by isometry, and using Lemma 2.3.1

$$\lim_{n\to\infty} \tilde{\rho}'(g(\tilde{x}_n), g(\tilde{y}_n)) = \lim_{n\to\infty} \tilde{\rho}(\tilde{x}_n, \tilde{y}_n) = 0,$$

and, using Lemma 2.3.1 again, it follows that

$$\lim_{n\to\infty} g(\tilde{x}_n) = \lim_{n\to\infty} g(\tilde{y}_n).$$

Although $g: \tilde{Y} \to \tilde{Y}'$ is bijective, an arbitrary extension of g will not necessarily also be bijective; it is now shown that the particular extension defined above is indeed bijective.

First let $\tilde{x}, \tilde{y} \in \tilde{X}$ and let $(\tilde{x}_n)$, $(\tilde{y}_n)$ be sequences in $\tilde{Y}$ which converge to $\tilde{x}, \tilde{y}$ respectively. Then

$$\begin{aligned}\tilde{\rho}(\tilde{x}, \tilde{y}) = \lim_{n\to\infty} \tilde{\rho}(\tilde{x}_n, \tilde{y}_n) &= \lim_{n\to\infty} \tilde{\rho}'(g(\tilde{x}_n), g(\tilde{y}_n)) \\ &= \tilde{\rho}'(G(\tilde{x}), G(\tilde{y})),\end{aligned}$$

so

$$\tilde{\rho}(\tilde{x}, \tilde{y}) = \tilde{\rho}'(G(\tilde{x}), G(\tilde{y})). \tag{4.5.1}$$

Now suppose that there exist $\tilde{x}, \tilde{y}$ in $\tilde{X}$ such that $G(\tilde{x}) = G(\tilde{y})$; then from (4.5.1) it follows that $\tilde{x} = \tilde{y}$, so G is injective.

To show that G is surjective let $\tilde{x}'$ be any point of $\tilde{X}'$ and let $(\tilde{x}_n')$ be an infinite sequence in $\tilde{Y}'$ which converges to $\tilde{x}'$. Then $(g^{-1}(\tilde{x}_n'))$ is a fundamental sequence of $(\tilde{Y}, \tilde{\rho})$ which therefore has a limit $\tilde{x}$ (say) in $\tilde{X}$. Using (4.5.1) we have

$$\tilde{\rho}(g^{-1}(\tilde{x}_n'), \tilde{x}) = \tilde{\rho}'(\tilde{x}_n', G(\tilde{x}));$$

on letting $n \to \infty$ it follows that $\tilde{\rho}'(\tilde{x}', G(\tilde{x})) = 0$. Hence $\tilde{x}' = G(\tilde{x})$ and G is surjective.

From (4.5.1) it now follows that $G:(\tilde{X}, \tilde{\rho}) \to (\tilde{X}', \tilde{\rho}')$ is an isometry.

4.6 The contraction mapping theorem

The remainder of the present chapter is devoted to this powerful result concerning complete metric spaces, and some applications of it.

First, two terms are defined.

DEFINITION 4.6.1. Let f be a mapping of a non-empty set X into itself. If there exists x' in X such that $f(x') = x'$, then x' is called a *fixed point* of f.

DEFINITION 4.6.2. Let f be a mapping of a metric space (X, ρ) into itself. Then f is called a *contraction mapping* if there exists λ such that, for all x, y in X,

$$\rho(f(x), f(y)) \leqslant \lambda\rho(x, y)$$

where $0 \leqslant \lambda < 1$.

It is stressed that

(i) λ is strictly less than 1, and

(ii) λ is independent of x, y in X.

The following result can be deduced immediately.

LEMMA 4.6.1. *Let (X, ρ) be a metric space and $f:(X, \rho) \to (X, \rho)$ be a contraction mapping; then f is uniformly continuous over X.*

The basic result, known as the contraction mapping theorem or Banach's fixed point theorem, is now established.

THEOREM 4.6.1. *Let (X, ρ) be a complete metric space, and $f:(X, \rho) \to (X, \rho)$ be a contraction mapping; then f has exactly one fixed point.*

Remark. The essence of the proof depends on the fact that if there exists a fixed point x', then the distance of any other point x from x' is reduced by f. Repeated applications of the mapping f shrinks this distance further, until in the limit it is zero; that is, writing

$$f^{(n)}(x) = f(f^{(n-1)}(x)), \quad n = 2, 3, \ldots, \quad f^{(1)}(x) = f(x),$$

then
$$\lim_{n\to\infty} f^{(n)}(x) = x'.$$

Therefore, in order to prove the existence of such an x', we look at the sequence $(f^{(n)}(x_0))$ for some x_0 in X, show that this does have a limit (without assuming the existence of a fixed point) and show finally that this limit is indeed the (unique) fixed point of f.

Proof. Let x_0 be any point of X, and define the sequence (x_n) in (X, ρ) by $x_n = f^{(n)}(x_0)$, $n = 1, 2, \ldots$; thus $x_n = f(x_{n-1})$.

First we show that this sequence is fundamental. For

$$\begin{aligned}\rho(x_n, x_{n+1}) = \rho(f(x_{n-1}), f(x_n)) &\leqslant \lambda\rho(x_{n-1}, x_n)\\ &\leqslant \lambda^2\rho(x_{n-2}, x_{n-1})\\ &\quad \cdot \quad \cdot \quad \cdot\\ &\leqslant \lambda^n\rho(x_0, x_1) = \lambda^n a,\end{aligned}$$

say. Let m, n be any integers such that $m < n$; then

$$\begin{aligned}\rho(x_m, x_n) &\leqslant \rho(x_m, x_{m+1}) + \rho(x_{m+1}, x_{m+2}) + \ldots + \rho(x_{n-1}, x_n)\\ &\leqslant \lambda^m a + \lambda^{m+1} a + \ldots + \lambda^{n-1} a\\ &\leqslant \frac{\lambda^m a}{1-\lambda},\end{aligned}$$

since $0 \leqslant \lambda < 1$. Therefore it follows that $\rho(x_m, x_n) \to 0$ as $m \to \infty$ (so $n \to \infty$ also), and so (x_n) is a fundamental sequence.

Since (X, ρ) is complete, (x_n) converges to some element x' of X; this is a fixed point of f. For letting $n \to \infty$ in $x_{n+1} = f(x_n)$ and using that f is continuous (see Lemma 4.6.1) it follows immediately that $x' = f(x')$.

Now let x'' be any fixed point of f; then

$$\rho(x', x'') = \rho(f(x'), f(x'')) \leqslant \lambda\rho(x', x'').$$

Since $0 \leqslant \lambda < 1$ it follows that $\rho(x', x'') = 0$, so $x' = x''$; thus the fixed point of f is unique.

The following result gives a useful extension of the above theorem.

COROLLARY 4.6.1. *Let (X, ρ) be a complete metric space and $f:(X, \rho) \to (X, \rho)$ be a mapping such that, for some positive integer p, $f^{(p)}$ is a contraction mapping, say*

$$\rho(f^{(p)}(x), f^{(p)}(y)) \leqslant \lambda\rho(x, y)$$

where $0 \leqslant \lambda < 1$; then f has exactly one fixed point.

Proof. By Theorem 4.6.1 there is a unique point x' such that $f^{(p)}(x') = x'$. Then

$$\rho(f(x'), x') = \rho(f^{(p+1)}(x'), f^{(p)}(x')) \leqslant \lambda\rho(f(x'), x')$$

so that, as above, we have $f(x') = x'$. Thus x' is a fixed point of f; by a simple contradiction argument it follows that f has only one fixed point.

It is straightforward to establish a bound for the rate at which the sequence (x_n), defined in the proof of Theorem 4.6.1, converges to the fixed point x' of f.

By the triangle inequality, for any integer $m > n$,

$$\begin{aligned}\rho(x_n, x') &\leqslant \rho(x_n, x_{n+1}) + \rho(x_{n+1}, x_{n+2}) + \ldots + \rho(x_m, x') \\ &\leqslant (\lambda^n + \lambda^{n+1} + \ldots + \lambda^{m-1})a + \rho(x_m, x') \\ &\leqslant \frac{\lambda^n}{1-\lambda} a + \rho(x_m, x').\end{aligned}$$

Since this holds for all m ($> n$), letting $m \to \infty$ it follows that

$$\rho(x_n, x') \leqslant \frac{\lambda^n}{1-\lambda} \rho(x_1, x_0);$$

this is the required expression.

We conclude this section with two remarks.

(i) The reader should observe that the proof of Banach's theorem is *constructive*; that is, the existence of a fixed point is established by constructing the point (as the limit of a sequence of points tending to the fixed point). The applications of the next section illustrate the use of this.

(ii) Banach's theorem is a particular example of a very wide class of similar but more general results concerning the existence of fixed points; such theorems play an important part in modern mathematics, even when the proofs are not constructive.

EXERCISES 4.6

1. (i) Let $f: [a, b] \to \mathsf{R}$ be differentiable over $[a, b]$. Show that f is a contraction mapping if and only if there exists a constant λ such that $|f'(x)| \leqslant \lambda < 1$ for all x in $[a, b]$.

(ii) By means of the following example show that it is not possible to omit the requirement of completeness in Theorem 4.6.1. Let $X = [1,2] \cap \mathrm{Q}$, ρ be the Euclidean metric and

$$f(x) = -\tfrac{1}{4}(x^2-2)+x.$$

2. By means of the following example show that, in general, it is not possible to replace the condition

$$\rho(f(x), f(y)) \leqslant \lambda\rho(x, y), \quad 0 \leqslant \lambda < 1$$

in Theorem 4.6.1 by

$$\rho(f(x), f(y)) < \rho(x, y)$$

(for all distinct x, y in X). Let $X = \mathrm{R}$, ρ be the Euclidean metric and

$$f(x) = \tfrac{1}{2}+x-\tan^{-1} x$$

(taking the value of $\tan^{-1}$ lying in $(-\frac{1}{2}\pi, \frac{1}{2}\pi)$); in this case show that f has no fixed points. [See, however, Exercises 5.7.7, 5.8.6.]

3. Let (X, ρ) be a complete metric space and let (T, σ) be another metric space. Let $X' = X \times T$ and

$$\rho(x_1', x_2') = \max\{\rho(x_1, x_2), \sigma(\tau_1, \tau_2)\}$$

where $x_1' = (x_1, \tau_1)$, $x_2' = (x_2, \tau_2)$. Let $f:(X', \rho') \to (X, \rho)$ be continuous and satisfy the contraction condition

$$\rho(f(x, \tau), f(y, \tau)) \leqslant \lambda\rho(x, y)$$

for all τ in T and all x, y in X, where λ is a constant such that $0 \leqslant \lambda < 1$.

Let $g: T \to X$ be defined by $f(g(\tau), \tau) = g(\tau)$; the existence of such a function g follows from Banach's theorem. Show that

$$\rho(g(\tau), g(\tau')) \leqslant \frac{\sigma(\tau, \tau')}{1-\lambda}$$

for all τ, τ' in T; deduce that $g:(T, \sigma) \to (X, \rho)$ is continuous.

[This result shows, roughly speaking, that if the contraction mapping in Banach's theorem depends continuously on some parameter τ, then the corresponding unique fixed point also depends continuously on τ.]

4. Let $I = [0, 1]$; associate the usual supremum metric ρ with $\mathscr{C}(I)$. Define $F:(\mathscr{C}(I), \rho) \to (\mathscr{C}(I), \rho)$ by

$$\{F(f)\}(x) = \int_0^x f(t)\,dt$$

for all f in $\mathscr{C}(I)$. Show that

(i) $\{F(f)\}(x)-\{F(g)\}(x) \leqslant x.\rho(f, g)$,

(ii) $\{F^{(2)}(f)\}(x)-\{F^{(2)}(g)\}(x) \leqslant \frac{1}{2}x^2.\rho(f, g)$,

for all f, g in $\mathscr{C}(I)$, and deduce that $F^{(2)}$ is a contraction mapping. Show, however, that F is not a contraction mapping.

4.7 Simple applications of the contraction mapping theorem

First it is necessary to define some terms.

DEFINITION 4.7.1. A function $f: X(\subseteq \mathsf{R}) \to \mathsf{R}$ is said to satisfy a *Lipschitz condition of order* α (> 0) *at the point* x_0 (in X) if there exists $M > 0$ and $\delta > 0$ for which

$$|f(x)-f(x_0)| \leqslant M|x-x_0|^{\alpha}$$

for all x in X such that $|x-x_0| < \delta$.

Furthermore f is said to satisfy a *Lipschitz condition of order* α (> 0) *throughout* (or *on*) X if there exists $M > 0$ such that

$$|f(x)-f(y)| \leqslant M|x-y|^{\alpha}$$

for all x, y in X. This is sometimes written $f \in \text{Lip}_{\alpha}(X)$.

It is easily seen that if f satisfies a Lipschitz condition of any order α (> 0), either at a point x_0, or throughout X, then f is continuous there. The most important case is that of $\alpha = 1$, and is the only one which will concern us.

Some elementary applications of the contraction mapping theorem are now given.

I. *Solution of the real variable equation* $f(x) = x$.

Let $I = [a, b]$. Suppose that f is a function which maps I into I, and satisfies on I the Lipschitz condition

$$|f(x)-f(y)| \leqslant K|x-y|, \quad K < 1. \tag{4.7.1}$$

From (4.7.1) it follows that f is continuous, so the function $g: I \to I$ defined by $g(x) = f(x)-x$ is also continuous. Now $f(a) \in I$ so $f(a) > a$ and $g(a) > 0$; likewise $g(b) < 0$. Therefore by the intermediate value theorem $g(x) = 0$, and hence $f(x) = x$, has a solution, x' say, in I. Moreover from (4.7.1) if $x > y$ we have

$$f(x)-f(y) < x-y,$$

so $g(x) < g(y)$ and thus g is increasing on I; hence x' is unique.

Then the proof of the contraction mapping theorem provides a numerical method for calculating x' approximately. For, from (4.7.1) it follows that f is a contraction mapping over (I, d); since I is a closed subset of (R, d), which is complete, (I, d) is complete. Hence by the iterative process used in the proof of the contraction mapping theorem, x' can be calculated approximately from

$$x' = \lim_{n\to\infty} f^{(n)}(x_0),$$

where x_0 is any point in I.

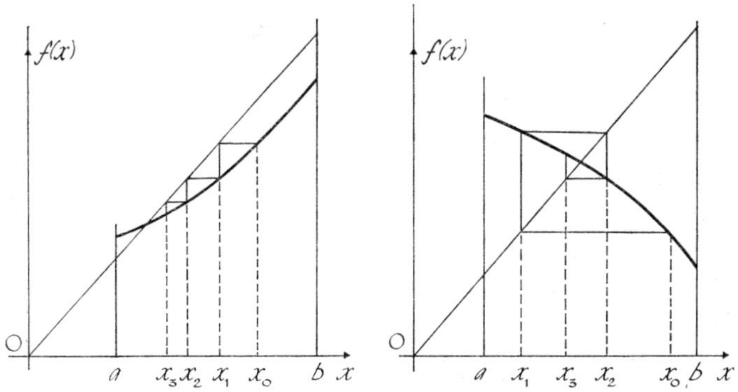

The diagrams illustrate a geometrical construction for the sequence (x_n) when f is strictly increasing and f is strictly decreasing respectively. It is left to the reader to examine the form of the construction.

II. *Solution of the real variable equation* $F(x) = 0$.

The above method can be readily adapted to establish the existence of unique solutions of $F(x) = 0$ under suitable conditions.

Let $I = [a, b]$; suppose that $F: I \to \mathsf{R}$ is a differentiable function for which

(i) $F(a)F(b) < 0$,

(ii) there exist constants m, M such that $0 < m \leqslant F'(x) \leqslant M$ on I.

Then by the intermediate value theorem, $F(x) = 0$ has a solution in I; moreover since F is strictly increasing over I this solution must be unique. Again the proof of the contraction mapping theorem provides a numerical method for calculating an approximate solution.

D

To see this, set $f(x) = x-\lambda F(x)$; then $F(x) = 0$ if and only if $f(x) = x$. From (ii) it follows that

$$1-\lambda M \leqslant f'(x) \leqslant 1-\lambda m.$$

Now choose λ such that $0 < \lambda < M^{-1}$, so $1-\lambda M > 0$. Then

$$0 < f'(x) \leqslant 1-\lambda m; \tag{4.7.2}$$

hence f is strictly increasing. Since $f(a)-a = -\lambda F(a) > 0$, it follows that $f(a) > a$; similarly $f(b) < b$. Hence $a < f(x) < b$ for all x in I, so f maps I into I. Lastly, by integrating (4.7.2) it is easily seen that $f:(I, \mathsf{d}) \to (I, \mathsf{d})$ is a contraction.

III. *Solution of a finite system of linear equations.*

Let† $b^{\mathrm{T}} \in \mathsf{R}^m$, $A = (a_{ij})$ be a real $m \times m$ matrix and $f: \mathsf{R}^m \to \mathsf{R}^m$ be the mapping defined by $f(x) = Ax+b$, where $x^{\mathrm{T}} \in \mathsf{R}^m$. We seek conditions for $f(x) = x$ to possess a unique solution, and which permit us to use the iterative procedure of the contraction mapping theorem to estimate the solution.

The first question to be settled is 'under what conditions is f a contraction mapping?' This of course depends on the matrix A, and also on the metric which is associated with the set R^m; we associate the Euclidean metric with R^m.

Let $x = (x_1, \ldots, x_m)$, $y = (y_1, \ldots, y_m)$; then the ith component of $f(x)-f(y)$ is

$$\textstyle\sum_j \{a_{ij}(x_j-y_j)\}.$$

Hence

$$\begin{aligned} \mathsf{d}^2(f(x), f(y)) &= \textstyle\sum_i \{\sum_j a_{ij}(x_j-y_j)\}^2 \\ &\leqslant \textstyle\sum_i \{\sum_j a_{ij}^2 \sum_j (x_j-y_j)^2\}, \end{aligned}$$

by Cauchy's inequality, so

$$\mathsf{d}^2(f(x), f(y)) \leqslant (\textstyle\sum_i \sum_j a_{ij}^2)\, \mathsf{d}^2(x, y).$$

Therefore, if

$$\textstyle\sum_i \sum_j a_{ij}^2 < 1, \tag{4.7.3}$$

† b^{T} denotes the transpose of b; thus b is a real m-tuple written as a column vector.

then f is a contraction mapping in $(\mathsf{R}^m, \mathsf{d})$. (Although (4.7.3) is a sufficient condition for f to be a contraction mapping in $(\mathsf{R}^m, \mathsf{d})$, in fact it can be shown to be not a necessary one.)

Thus if condition (4.7.3) is satisfied by the matrix A, then there exists a unique solution of $f(x) = x$ in R^m which can be calculated, approximately, by the now familiar process.

If we associate a different metric with R^m we would obtain different conditions on A. For details see Kolmogorov and Fomin (1957) §14, or Copson (1968) §74. The latter, in §75, also discusses infinite systems of linear equations.

EXERCISES 4.7

1. Verify that the function $f: \mathsf{R}^2 \to \mathsf{R}$ defined by $f(x, y) = xy^2$ satisfies a Lipschitz condition of the form

$$|f(x, y_1) - f(x, y_2)| \leqslant K|y_1 - y_2|$$

on the rectangle $[-1, 1] \times [-1, 1]$, but satisfies no such condition on the strip $[-1, 1] \times (-\infty, \infty)$.

2. If the function $f: \mathsf{R} \to \mathsf{R}$ satisfies a Lipschitz condition of order α at x_0, show that if $\alpha > 0$, then f is continuous at x_0, while if $\alpha > 1$, then f is differentiable at x_0. Give an example of a function that satisfies a Lipschitz condition of order 1 at some point x_0, but is not differentiable at x_0.

3. In R^2 let $X = [x_0 - a, x_0 + a] \times [y_0 - b, y_0 + b]$, and let $f: X \to \mathsf{R}$ be such that $\partial f/\partial y$ exists and $|\partial f/\partial y| \leqslant K$ throughout X, for some positive constant K. Show that f satisfies a Lipschitz condition

$$|f(x, y_1) - f(x, y_2)| \leqslant K|y_1 - y_2|$$

(of order 1) for all y_1, y_2 in $[y_0 - b, y_0 + b]$.

Also show that $[y_0 - b, y_0 + b]$ may be replaced, throughout, by $(-\infty, \infty)$.

4. Associate the usual supremum metric ρ with $\mathscr{C}(I)$ where $I = [a, b]$. Let M_K denote the subset of $\mathscr{C}(I)$ of all functions f which satisfy the Lipschitz condition

$$|f(t) - f(t')| \leqslant K|t - t'|$$

for all t, t' in I, and let D_K denote the subset of all functions f which are differentiable over I and such that $|f'(t)| \leqslant K$ there. Show that M_K is a closed subset of $(\mathscr{C}(I), \rho)$, and that $M_K = \bar{D}_K$.

4.8 Picard's theorem

The contraction mapping theorem will now be used to prove an important result concerning the existence of a unique solution to a first order ordinary differential equation satisfying certain conditions. This equation is first shown to be equivalent to an integral equation.

LEMMA 4.8.1. *Let* $I = [x_0-a, x_0+a]$ *and* $S = I\times \mathsf{R}$; *let* $f:S\to \mathsf{R}$ *be continuous.*

Then any solution† y *over* I *of the differential equation*

$$y' = f(x, y) \qquad (y' = dy/dx) \tag{4.8.1}$$

such that $y(x_0) = y_0$ *will also be a solution*‡ *over* I *of the integral equation*

$$y(x) = y_0+\int_{x_0}^{x} f(t, y(t))\,dt. \tag{4.8.2}$$

Conversely any solution y *over* I *of the integral equation* (4.8.2) *will also be a solution over* I *of the differential equation* (4.8.1) *subject to* $y(x_0) = y_0$.

Proof. We disregard for the moment all question of rigour. If y is a solution of (4.8.1) over I satisfying $y(x_0) = y_0$, then by integrating (4.8.1) it follows that

$$\int_{x_0}^{x} y'(t)\,dt = \int_{x_0}^{x} f(t, y(t))\,dt,$$

and since

$$\int_{x_0}^{x} y'(t)\,dt = y(x)-y(x_0), \tag{4.8.3}$$

we obtain (4.8.2).

Furthermore by differentiating (4.8.2), since

$$\frac{d}{dx}\int_{x_0}^{x} f(t, y(t))\,dt = f(x, y(x)), \tag{4.8.4}$$

it follows that any solution y of (4.8.2) over I is also a solution of (4.8.1) such that $y(x_0) = y_0$.

To make the above argument rigorous it is necessary to justify (4.8.3) and (4.8.4).

† A solution of (4.8.1) is any differentiable function $y: I\to \mathsf{R}$ such that $y'(x) = f(x, y(x))$ for all x in I.

‡ A solution of (4.8.2) is any continuous function $y: I\to \mathsf{R}$ such that (4.8.2) is satisfied for all x in I.

For (4.8.3) note first that since y is differentiable over I, it is continuous there. Then, since $y'(t) = f(t, y(t))$ and since $t \in I$ implies that $(t, y(t)) \in S$, it follows that the function $t \mapsto f(t, y(t))$, and hence y', is continuous over I. Therefore (4.8.3) follows (by Theorem 1.6.4).

For (4.8.4), if y is a solution of (4.8.2) over I, which implies that the function $t \mapsto f(t, y(t))$ is integrable over I, then y is continuous over I (by Theorem 1.6.3). Therefore $t \mapsto f(t, y(t))$ is continuous over I and so (by Theorem 1.6.3 again) the function

$$x \mapsto \int_{x_0}^{x} f(t, y(t))\,dt$$

is differentiable; hence (4.8.4) holds.

The basic theorem concerning the existence (and uniqueness) of solutions of $y' = f(x, y)$ may now be established. The original proof of this result was given by Picard; he did not appeal to the contraction mapping theorem, but instead constructed a sequence of approximate solutions which converges to a solution of the differential equation. This solution, then, has to be shown to be unique. The contraction mapping theorem gives both conclusions simultaneously.

THEOREM 4.8.1. *Let* $I = [x_0 - a, x_0 + a]$ *and* $S = I \times \mathsf{R}$; *let* $f: S \to \mathsf{R}$ *be continuous. Suppose also that f satisfies a Lipschitz condition* (*of order* 1) *in y which is uniform in x over I; that is*

$$|f(x, y_1) - f(x, y_2)| \leqslant K|y_1 - y_2|$$

for all y_1, y_2 *in* R, *where K is independent of x in I.*

Then the differential equation

$$y' = f(x, y)$$

possesses a unique solution through† (x_0, y_0) *over I.*

Proof. In view of Lemma 4.8.1, we consider the integral equation

$$y(x) = y_0 + \int_{x_0}^{x} f(t, y(t))\,dt \tag{4.8.5}$$

where $x \in I$, and show that this possesses a unique solution over I.

We consider first the half interval $I_0 = [x_0, x_0 + a]$. Associate with $\mathscr{C}(I_0)$, the set of all real-valued functions defined and continuous over I_0, the metric ρ given by

$$\rho(f, g) = \sup_{x \in I_0} e^{-k(x - x_0)}|f(x) - g(x)|, \tag{4.8.6}$$

† By a solution through (x_0, y_0) we mean a solution y such that $y(x_0) = y_0$.

where k is a positive constant to be chosen later; it is easily verified that (4.8.6) does indeed satisfy the metric space axioms, and (by imitating the arguments of §4.2(viii)) that $(\mathscr{C}(I_0), \rho)$ is complete. (See also Exercise 4.8.1 below.) The reason for including the exponential factor $e^{-k(x-x_0)}$ in the definition of ρ will transpire shortly.

Define a mapping F on $(\mathscr{C}(I_0), \rho)$ by

$$\{F(y)\}(x) = y_0 + \int_{x_0}^{x} f(t, y(t))\, dt$$

where $y \in \mathscr{C}(I_0)$; then the function $t \mapsto f(t, y(t))$ is continuous over I_0, so $F(y)$ is continuous there and thus F maps $(\mathscr{C}(I_0), \rho)$ into itself. It will now be shown that F is a contraction mapping on $(\mathscr{C}(I_0), \rho)$, when k is chosen suitably.

For any y_1, y_2 in $\mathscr{C}(I_0)$

$$\begin{aligned} \left|\{F(y_1)\}(x) - \{F(y_2)\}(x)\right| &\leqslant \int_{x_0}^{x} \left|f(t, y_1(t)) - f(t, y_2(t))\right| dt \\ &\leqslant K \int_{x_0}^{x} \left|y_1(t) - y_2(t)\right| dt, \end{aligned}$$

and thus

$$\begin{aligned} e^{-k(x-x_0)}&\left|\{F(y_1)\}(x) - \{F(y_2)\}(x)\right| \\ &\leqslant K \int_{x_0}^{x} e^{-k(x-x_0)} \left|y_1(t) - y_2(t)\right| dt \\ &= K \int_{x_0}^{x} e^{-k(x-t)} \{e^{-k(t-x_0)} \left|y_1(t) - y_2(t)\right|\} dt \\ &\leqslant K\rho(y_1, y_2) \int_{x_0}^{x} e^{-k(x-t)}\, dt \\ &\leqslant \frac{K}{k} \cdot \rho(y_1, y_2). \end{aligned}$$

Now choose $k = 2K$; taking suprema it follows that

$$\rho(F(y_1), F(y_2)) \leqslant \tfrac{1}{2}\rho(y_1, y_2).$$

Thus it is seen how the exponential factor of $e^{-k(x-x_0)}$ in (4.8.6) ensures that F is a contraction mapping of $(\mathscr{C}(I_0), \rho)$ into itself.

Hence, by Banach's theorem, there exists a unique element g of $\mathscr{C}(I_0)$ such that $F(g) = g$, that is, which satisfies the integral equation (4.8.5). Likewise there is a unique continuous function over $[x_0 - a, x_0]$ which satisfies (4.8.5), and hence we have a unique solution over the whole interval $[x_0 - a, x_0 + a]$.

This completes the proof.

The neat device of introducing an exponential weighting factor in the definition of the metric ρ defined in (4.8.6) is due to Bielecki (1956).

In both Lemma 4.8.1 and Theorem 4.8.1, the function f was assumed to be continuous over the strip $[x_0-a, x_0+a]\times(-\infty, \infty)$; by making slight changes in the statement and proof of these results it is sufficient to assume continuity of f over a rectangle of the form

$$[x_0-a, x_0+a]\times[y_0-b, y_0+b].$$

Furthermore it is possible to extend the above theorem to more general situations, in particular to higher order equations of a similar form, and to equations involving more than one variable. All of these extensions fall, however, more appropriately within the realm of the theory of differential equations.

The contraction mapping theorem was applied above to a certain integral equation; it can be applied, in a similar manner, to a number of types of integral equations. Some examples of these are given in the exercises.

Exercises 4.8

1. Let ρ be the metric defined on $\mathscr{C}(I_0)$ by (4.8.6) and let ρ' be the usual supremum metric on $\mathscr{C}(I_0)$. Show that ρ, ρ' are uniformly equivalent; using Exercise 4.1.5 deduce that $(\mathscr{C}(I_0), \rho)$ is complete.

2. Find a sequence (y_n) of functions y_n such that y_n converges to the solution of the integral equation (4.8.5).

3. Let $I = [x_0-a, x_0+a]$ and $S = I\times \mathsf{R}$; let $f:S\to \mathsf{R}$ be continuous. Show that y is a solution over I of the differential equation

$$\frac{d^2y}{dx^2} = f(x, y)$$

such that $y(x_0) = y_0$, $y'(x_0) = y_1$ if and only if it is a solution over I of the integral equation

$$y(x) = y_0+(x-x_0)y_1+\int_{x_0}^{x}(x-t)f(t, y(t))\,dt.$$

If f also satisfies a Lipschitz condition (of order 1) in y which is uniform in x over I, prove that the given differential equation possesses a unique solution such that $y(x_0) = y_0$, $y'(x_0) = y_1$.

4.† Let $I = [a, b]$ and $X = I \times I$. Suppose that the functions $K: X \to \mathsf{R}$ and $\phi: I \to \mathsf{R}$ are continuous; let

$$M = \sup_{(x,y) \in X} |K(x, y)|.$$

Prove that the integral equation

$$f(x) = \lambda \int_a^b K(x, y) f(y)\, dy + \phi(x),$$

where λ is an arbitrary real parameter, will have a unique solution which is continuous, provided $|\lambda| < 1/M(b-a)$.

[In this case it will be seen that the metric defined in (4.8.6) holds no advantage over the usual supremum metric.]

5.‡ Let $I = [a, b]$ and let $X = I \times I \times \mathsf{R}$. Suppose that the function $K: X \to \mathsf{R}$ is continuous and satisfies the Lipschitz condition

$$|K(x, y, z_1) - K(x, y, z_2)| \leqslant M|z_1 - z_2|$$

for all z_1, z_2 in R, where M is independent of x, y in I; suppose also that the function $\phi: I \to \mathsf{R}$ is continuous.

Prove that the integral equation

$$f(x) = \lambda \int_a^b K(x, y, f(y))\, dy + \phi(x),$$

where λ is an arbitrary real parameter, will have a unique solution which is continuous, provided $|\lambda| < 1/M(b-a)$.

[Again observe that the metric defined in (4.8.6) holds no advantage over the usual supremum metric.]

6. Let $I = [a, b]$ and $X = I \times I$. Suppose that the functions $K: X \to \mathsf{R}$ and $\phi: I \to \mathsf{R}$ are continuous.

Prove that the integral equation

$$f(x) = \lambda \int_a^x K(x, y) f(y)\, dy + \phi(x),$$

where λ is an arbitrary real parameter, will have a unique solution which is continuous, for any λ.

[Does the metric defined by (4.8.6) hold any advantage over the usual metric in this case?]

† (ii) of Theorem 1.6.1 has the following two dimensional analogue (which is required in Exercise 4.8.4). If $J = [a, b] \times [c, d]$ and if $F: J \to \mathsf{R}$ is continuous over J, then the set $F(J)$ is bounded. The reader may also wish to use in Exercises 4.8.4, 4.8.5, the fact that, under the same hypothesis, F is uniformly continuous over J. These assertions can be established in the same way as (i), (ii) of Theorem 1.6.1; alternatively they will follow as special cases of Theorems 5.8.1, 5.8.4.

‡ See footnote for Exercise 4.8.4.

4.9 An implicit function theorem

If $f: \mathsf{R}^2 \to \mathsf{R}$ is defined by $f(x, y) = x^2 - y^2$, then there exist many functions $y: \mathsf{R} \to \mathsf{R}$ which satisfy $f(x, y) = 0$; for example $y_1(x) = x$, $y_2(x) = -x$, both of which are continuous, but there is also

$$y_3(x) = \begin{cases} x, \text{ if } x \text{ rational} \\ -x, \text{ if } x \text{ irrational} \end{cases}$$

which is not continuous. On the other hand, if f is defined by $f(x, y) = x^2 + y^2 + 1$ then there is no real solution to $f(x, y) = 0$.

These simple examples introduce the following problem of analysis. Given a function $f: X(\subseteq \mathsf{R}^2) \to \mathsf{R}$ does there exist a function $g: Y(\subseteq \mathsf{R}) \to \mathsf{R}$ such that

$$f(x, g(x)) = 0 \tag{4.9.1}$$

for all x in Y? Will g be unique? Furthermore, if f is a suitably smooth function, say continuous over X, what can we say about the smoothness of g?

There are various results giving necessary conditions on f that $f(x, y) = 0$ is 'solvable' in the sense that there exists a function g for which (4.9.1) is true; one of the most basic of such results is given by Apostol (1957), p. 147. Here we give a closely related result which may be established by using the contraction mapping theorem.

THEOREM 4.9.1. *Let $I = [a, b]$ and $S = I \times \mathsf{R}$; let $f: S \to \mathsf{R}$ be continuous. Suppose also that f satisfies the condition*

$$0 < m \leqslant \frac{f(x, y) - f(x, z)}{y - z} \leqslant M \tag{4.9.2}$$

for all distinct y, z in R.

Then there exists a unique function g in $\mathscr{C}(I)$ such that $f(x, g(x)) = 0$ for all x in I.

Proof. Define a mapping $F: \mathscr{C}(I) \to \mathscr{C}(I)$ by

$$\{F(g)\}(x) = g(x) - \frac{2}{M+m} f(x, g(x))$$

where $g \in \mathscr{C}(I)$; associate the usual supremum metric ρ with $\mathscr{C}(I)$, so that $(\mathscr{C}(I), \rho)$ is complete. It will be shown that F is a contraction mapping.

First observe that

$$\{F(g)\}(x)-\{F(h)\}(x) = g(x)-h(x)-\frac{2}{M+m}\{f(x,g(x))-f(x,h(x))\}.$$

From (4.9.2) it follows that

$$-\frac{M-m}{M+m} \leqslant 1-\frac{2}{M+m}\cdot\frac{f(x,y)-f(x,z)}{y-z} \leqslant \frac{M-m}{M+m},$$

and therefore

$$\begin{aligned}\rho(F(g), F(h)) &= \sup_x \left|\{F(g)\}(x)-\{F(h)\}(x)\right| \\ &\leqslant \sup_x \left|g(x)-h(x)\right|\cdot\frac{M-m}{M+m} \\ &= \frac{M-m}{M+m}\cdot\rho(g, h) \\ &= \lambda\rho(g, h),\end{aligned}$$

say. Since $M \geqslant m > 0$ it follows that $0 \leqslant \lambda < 1$, so F is a contraction mapping on $(\mathscr{C}(I), \rho)$.

The latter is complete so there exists a unique element g of $\mathscr{C}(I)$ such that $F(g) = g$, that is, such that $f(x, g(x)) = 0$ for all x in I.

5: COMPACTNESS

5.1 Introduction

In §4.1 the reader was reminded of the Cauchy principle of convergence; since this is such a fundamental result concerning convergence in the real number system, the question arose as to whether or not a generalization of the result held in any metric space. The answer to this was easily found to be that such a generalization did not necessarily hold. We therefore gave a special name to those spaces in which it did hold (namely complete metric spaces) and proceeded to study them. In this chapter we take several other important theorems which hold on certain subsets of the real line. Each of the results holds only on closed bounded sets; this is not coincidental, for the results are essentially related. How these results may be generalized to arbitrary metric spaces will now be investigated; this is a lengthy process.

The introductory details are now given. For the remainder of this section it is to be assumed that the Euclidean metric is associated with R.

PROPOSITION 5.1.1 (The Heine-Borel theorem). *Let Y be any closed bounded set of* R. *Then every class of open sets of* R, *whose union covers Y, contains a finite subclass whose union also covers Y.*

Thus if $\{Y_\lambda : \lambda \in \Lambda\}$ is a class of subsets of R, each Y_λ being open, and such that

$$\bigcup_{\lambda \in \Lambda} Y_\lambda \supseteq Y,$$

Proposition 5.1.1 asserts that, if Y is bounded and closed, then there exists a finite number of open sets Y_λ, say $Y_{\lambda_1}, \ldots, Y_{\lambda_n}$, such that

$$Y_{\lambda_1} \cup \ldots \cup Y_{\lambda_n} \supseteq Y.$$

PROPOSITION 5.1.2 (The Bolzano-Weierstrass theorem for sets). *Every bounded infinite set Y of* R *possesses a limit point; if Y is closed, the limit point is in Y.*

PROPOSITION 5.1.3 (The Bolzano-Weierstrass theorem for sequences). *Every bounded sequence in* R *contains a convergent subsequence.*

Propositions 5.1.1, 5.1.2 may be proved by the bisection method (for an illustration of this method at work see the proof of Proposition 5.3.1); Proposition 5.1.3 may be deduced from Proposition 5.1.2. There also exist certain results of a converse nature; for example, a set Y possesses the 'open-covering' property described in Proposition 5.1.1 only if Y is closed and bounded. Similar comments apply to Proposition 5.1.2, and less directly to Proposition 5.1.3. These can be pieced together to form the following result. The reader should not be distressed if he has not previously encountered all the implications of Proposition 5.1.4; they will all follow as special cases of Theorem 5.6.1.

PROPOSITION 5.1.4. *Let* $Y \subseteq \mathsf{R}$; *then the following statements are equivalent*:

(i) *every class of open sets of* R *whose union covers* Y *contains a finite subclass whose union covers* Y;

(ii) Y *is closed and bounded*;

(iii) *every infinite set in* Y *possesses a limit point in* Y;

(iv) *every sequence in* Y *contains a convergent subsequence whose limit is in* Y.

We now ask whether this result extends to an arbitrary metric space; the answer to this is given in §§5.2–5.6. In brief it is as follows. The natural generalizations of (i), (iii), (iv) are mutually equivalent in any metric space; however the natural generalization of (ii) is not equivalent to any of the generalizations of (i), (iii), (iv). This is illustrated by the following counter-example.

Consider the metric space (ℓ^p, ρ), defined in §2.2(vi); let Y be the subset of ℓ^p whose elements are $x^{(i)}, i \in \mathsf{N}$, where

$$x^{(i)} = (0, 0, \dots, 0, 1, 0, \dots),$$

the non-zero entry being in the ith place. Also let $x^{(0)} = (0, 0, 0, \dots)$; then $x^{(i)} \in S(x^{(0)}, 2)$ for all i, so Y is bounded. Now $\rho(x^{(m)}, x^{(n)}) = 2^{1/p}$ for all m, n $(m \neq n)$; therefore Y can contain no fundamental sequence, and so no convergent subsequence. Hence, by Theorem

2.5.1, Y has no limit points, and so is closed (recall that by Theorem 2.6.1 a set is closed if and only if it contains all its limit points).

Thus Y is a closed bounded infinite subset of (ℓ^p, ρ) which does not possess a limit point in Y. Moreover the sequence $(x^{(n)})$ in Y does not contain a convergent subsequence. Finally observe that the union of the class of open sets

$$\{S(x^{(n)}, 2^{1/(p+1)}) : n \in \mathsf{N}\}$$

covers Y but cannot contain a finite subclass whose union also covers Y. Therefore Y is closed and bounded but does not possess any of the properties corresponding to generalizations of (i), (iii), (iv) of Proposition 5.1.4.

Returning to the comments in the paragraph immediately following Proposition 5.1.4, we mention that although the natural generalization of (ii) is not equivalent to that of (i), (iii), (iv) there is, however, a rather different generalization of (ii) which is equivalent.

We shall simplify matters by considering, initially, only 'whole' metric spaces, and not subsets of metric spaces. We therefore study metric spaces which have the property that they satisfy the natural generalization of (i) of Proposition 5.1.4; that is, those metric spaces (X, ρ) for which every class of open subsets of X, whose union is X, contains a finite subclass whose union also is X. Such metric spaces will be called compact.

The natural generalizations of (iii) or (iv) could have alternatively been chosen for the definition of compactness—or even the generalization of (ii) when the correct form is known. In connection with this, L. V. Ahlfors in *Complex Analysis* (1966) writes on p. 60 as follows. 'There are several equivalent characterizations of compactness, and it is a matter of taste which one to choose as definition. Whatever we do the uninitiated reader will feel somewhat bewildered, for he will not be able to discern the purpose of the definition. This is not surprising, for it took a whole generation of mathematicians to agree on the best approach. The consensus of present opinion is that it is best to focus the attention on the different ways in which a given set can be covered by open sets.'

Of course one immediate and fundamental advantage of this definition of compactness is that it is meaningful in any topological space (see Definition 2.7.1).

Without further ado the formal definitions, theorems and rigorous proofs are given.

5.2 The definitions

DEFINITION 5.2.1. Let X be any non-empty set and $Y \subseteq X$; suppose that $\mathscr{A} = \{A_\lambda : \lambda \in \Lambda\}$ is a collection of subsets of X such that

$$Y \subseteq \bigcup_{\lambda \in \Lambda} A_\lambda.$$

Then $\mathscr{A}$ is called a *covering* of Y. If there exists a subset Λ' of Λ such that $\mathscr{A}' = \{A_\lambda : \lambda \in \Lambda'\}$ is also a covering of Y, then $\mathscr{A}'$ is called a *subcovering* of Y. If Λ is finite, then $\mathscr{A}$ is called a *finite covering*.

If (X, ρ) is a metric space, and $\mathscr{A} = \{A_\lambda : \lambda \in \Lambda\}$ is a covering of X such that A_λ is open for all λ in Λ, then $\mathscr{A}$ is called an *open covering* of (X, ρ).

DEFINITION 5.2.2. A metric space (X, ρ) is said to be *compact* if every open covering of (X, ρ) contains a finite subcovering.

The following paraphrase of this definition is due to Hermann Weyl. 'If a city is compact, it can be guarded by a finite number of arbitrarily near-sighted policemen.'

THEOREM 5.2.1. *If (X, ρ), (X', ρ') are homeomorphic, then (X, ρ) is compact if and only if (X', ρ') is compact.*

Proof. Let $f:(X, \rho) \to (X', \rho')$ be a homeomorphism. Suppose (X', ρ') is compact and let $\{Y_\lambda : \lambda \in \Lambda\}$ be an open covering of (X, ρ). Then clearly $\{f(Y_\lambda) : \lambda \in \Lambda\}$ is an open covering of (X', ρ'), so it contains a finite subcovering, say $\{f(Y_\lambda) : \lambda \in \Lambda'\}$. Hence $\{Y_\lambda : \lambda \in \Lambda'\}$ is a finite subcovering of (X, ρ) which is therefore compact.

It is necessary to introduce another concept before we can proceed with the discussion of compactness; metric spaces possessing this property will be called totally bounded. This property is, in general, more restrictive than ordinary boundedness, although it will later be seen that in the Euclidean space R^m the concepts are equivalent. The concept of total boundedness plays a simple but important role in connection with compactness.

DEFINITION 5.2.3. Let (X, ρ) be a metric space, and ε be an arbitrary positive number. Then a set $A \subseteq X$ is called an *ε-net* of (X, ρ) if, given any x in X, there exists at least one point a in A such that $\rho(a, x) < \varepsilon$. If the set A is finite, then A is called a *finite* ε-net of (X, ρ).

Thus, for example, in $(\mathrm{R}^2, \mathrm{d})$ the set of points of the form (m, n) where m, n are integers is a $(2^{-\frac{1}{2}}+\delta)$-net, for any $\delta > 0$.

Note that if A is an ε-net of (X, ρ), then

$$\bigcup_{a \in A} S(a, \varepsilon) = X.$$

DEFINITION 5.2.4. The metric space (X, ρ) is said to be *totally bounded* (or *precompact*) if, given any $\varepsilon > 0$, there exists a finite ε-net.

The following simple result is easily established.

LEMMA 5.2.1. *Any totally bounded metric space is bounded.*

Proof. If a metric space (X, ρ) is totally bounded, then there exists a finite 1-net, say $\{x_1, \ldots, x_n\}$; hence

$$X = \bigcup_{i=1}^{n} S(x_i, 1).$$

Since the union of a finite number of bounded sets is bounded, it follows that (X, ρ) is bounded.

EXERCISES 5.2

1. Find an open covering of $((0, 1), \mathrm{d})$ which does not contain a finite subcovering. Deduce that $((0, 1), \mathrm{d})$ and $([0, 1], \mathrm{d})$ are not homeomorphic.

2. Prove that any compact metric space is bounded.

3. If (X, ρ) is any metric space in which X is finite, show that (X, ρ) is compact.

4. A collection $\mathscr{Y} = \{Y_\lambda : \lambda \in \Lambda\}$ of subsets of some set X is said to have *empty intersection* if there is no point common to all the sets Y_λ; otherwise $\mathscr{Y}$ is said to have *non-empty intersection*. Furthermore $\mathscr{Y}$ is said to have the *finite intersection property* if every finite subcollection of subsets has non-empty intersection.

Let (X, ρ) be a metric space. Prove the following statements are equivalent:

(i) (X, ρ) is compact;

(ii) every collection of closed sets in (X, ρ) with empty intersection has a finite subcollection with empty intersection;

(iii) every collection of closed sets in (X, ρ) with the finite intersection property has non-empty intersection.

5. Using the previous exercise show that every nested sequence of non-empty closed sets in a compact metric space has non-empty intersection. [The converse of this result is also true; see Exercise 5.6.4.]

6. (i) Let (f_n) be a sequence of functions $f_n:(X, \rho) \to (\mathsf{R}, \mathsf{d})$ such that $f_n(x) \geqslant f_{n+1}(x)$ for $n = 1, 2, \ldots$ and all x in X; the sequence (f_n) is said to be *decreasing*. If each f_n is continuous over X and if, for some x_0 in X, $f_n(x_0) \to 0$ as $n \to \infty$, show that given any $\varepsilon > 0$ there exists N and δ (> 0) such that $f_n(x) < \varepsilon$ for all $n > N$ and all x in $S(x_0, \delta)$.

(ii) Hence establish the following theorem of Dini.

Let (X, ρ) be a compact metric space and let (f_n) be a decreasing sequence of continuous functions of (X, ρ) into (R, d) such that (f_n) converges pointwise to a function f. If f is continuous then the convergence is uniform. (There is, of course, a corresponding result if (f_n) is an increasing sequence.)

(iii) Give examples to show that none of the requirements: (X, ρ) is compact, f_n is decreasing (or alternatively increasing), f is continuous, in Dini's theorem may be omitted.

5.3 The first equivalent characterization of compactness

It follows, from the remark following Definition 5.2.3 and from Definition 5.2.4, that a metric space (X, ρ) is totally bounded if and only if, given any $\varepsilon > 0$, every covering of X by open spheres of radius ε contains a finite subcovering. This establishes the next result.

THEOREM 5.3.1. *Every compact metric space is totally bounded.*

We now establish another important property of compact metric spaces.

THEOREM 5.3.2. *Every compact metric space is complete.*

Proof. This will be by contradiction. Suppose, if possible, that (X, ρ) is a compact metric space which is not complete; let (x_n) be a fundamental sequence of (X, ρ) not having a limit in X.

Let $y \in X$; since (x_n) does not converge to y, there exists $\varepsilon_0 > 0$ such that

$$\rho(x_n, y) \geqslant 2\varepsilon_0 \tag{5.3.1}$$

for infinitely many values of n (see Lemma 1.7.1). Since (x_n) is a fundamental sequence, there exists N such that $\rho(x_m, x_n) < \varepsilon_0$ for all $m, n > N$. Choose a value of n, say n_0, for which (5.3.1) is satisfied and such that $n_0 > N$. Then since

$$\rho(x_{n_0}, y) \leqslant \rho(x_{n_0}, x_m) + \rho(x_m, y)$$

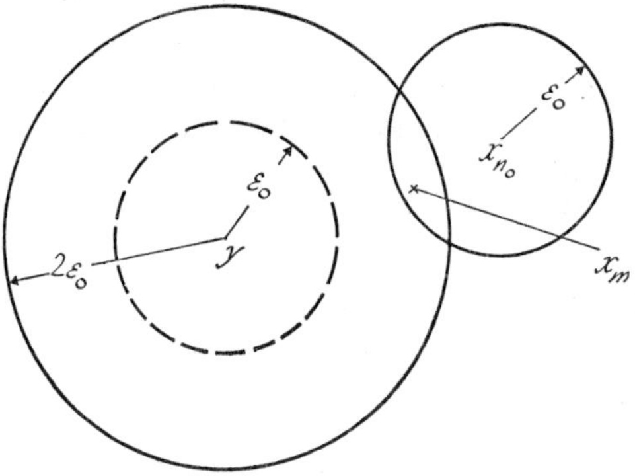

it follows that $\rho(x_m, y) > \varepsilon_0$ for all $m > N$. Thus $S(y, \varepsilon_0)$ contains x_n for only a finite number of values of n.

In this way, with each point y of X can be associated a sphere $S(y, \varepsilon_0)$, where ε_0 will depend on y; the collection of all such spheres forms an open covering of (X, ρ), which must contain, by hypothesis, a finite sub-covering. Each sphere contains x_n for only a finite number of values of n, so the finite subcovering (and therefore X) must contain x_n for only a finite number of values of n; but this is impossible, and so the theorem is established.

The reader's attention is drawn to the following point. The statement '$S(y, \varepsilon_0)$ contains x_n for only a finite number of values of n' is not equivalent to '$S(y, \varepsilon_0)$ contains only a finite number of points of the sequence (x_n)'. Of course the first statement implies the second, but the second does not imply the first.

It has therefore been proved that if a metric space is compact, then it is totally bounded and complete. The converse will now be established, namely that total boundedness and completeness together imply compactness; the proof of this is lengthy. The method given below is essentially an extension of the 'bisection' method; in order to make this proof easier to follow, a more straightforward example of a proof by the bisection method is given first.

PROPOSITION 5.3.1. *Let* $[a, b]$ *be any closed bounded interval of* R *(with Euclidean metric). Then any class of open subsets of* R *whose union contains* $[a, b]$, *possesses a finite subclass whose union also contains* $[a, b]$.

Proof. This will be by contradiction; suppose, if possible, that there exists at least one class of open subsets of R which is a covering of $[a, b]$, but which does not contain a finite subcovering of $[a, b]$. Let $\mathscr{A} = \{A_\lambda : \lambda \in \Lambda\}$ denote such a class.

Bisect $[a, b]$; then a finite number of the open sets A_λ cannot cover both half-intervals for otherwise $[a, b]$ would be covered by a finite number of the A_λ, which is impossible by hypothesis. Denote by $[a_1, b_1]$ that subinterval which is not covered by a finite number of the sets A_λ; if both half-intervals have this property, then for definiteness take the first to be $[a_1, b_1]$.

Bisect $[a_1, b_1]$; then as before at least one of the half-intervals, say $[a_2, b_2]$, is not covered by a finite number of the sets A_λ. Repeating this process, or more precisely by mathematical induction, for any positive integer n we can obtain an interval $[a_n, b_n]$ which is not covered by a finite number of the sets A_λ.

By the method of construction (a_n) is an increasing sequence which is bounded above (by b, for example) and (b_n) is a decreasing sequence which is bounded below (by a, for example); therefore these sequences must each possess a limit. Furthermore, since

$$b_n - a_n = \frac{b-a}{2^n},$$

these limits must be equal; denote their common value by ξ.

Clearly $\xi \in [a, b]$, so there exists λ_0 in Λ such that $\xi \in A_{\lambda_0}$; since A_{λ_0} is open there exists $\delta > 0$ for which $(\xi-\delta, \xi+\delta) \subseteq A_{\lambda_0}$. Now take n sufficiently large, with value n_0 say, such that

$$[a_{n_0}, b_{n_0}] \subseteq (\xi-\delta, \xi+\delta);$$

then

$$[a_{n_0}, b_{n_0}] \subseteq A_{\lambda_0} \in \mathscr{A}.$$

We have therefore found a finite subset of $\mathscr{A}$, the subset having only one member, namely A_{λ_0}, which covers $[a_{n_0}, b_{n_0}]$.

This gives the required contradiction, so the proposition has been established.

We now come to the main result.

Theorem 5.3.3. *If a metric space is complete and totally bounded, then it is compact.*

Proof. This will be by contradiction; suppose, if possible, that (X, ρ) is complete and totally bounded, but is not compact. Then there exists an open covering $\mathscr{A} = \{A_\lambda : \lambda \in \Lambda\}$ of X which contains no finite subcovering of X.

To aid understanding, the proof is divided into stages.

I. *To construct the sequence* $(S(x_n, \varepsilon_n))$.

Since (X, ρ) is totally bounded, it is bounded; hence for some r in R and some x_0 in X, $X = S(x_0, r)$ (note that $X \subseteq S(x_0, r)$ implies that $X = S(x_0, r)$). Let $\varepsilon_n = r/2^n$.

Since (X, ρ) is totally bounded, it can be covered by a finite number of spheres of radius ε_1. By our hypothesis at least one of these spheres, say $S(x_1, \varepsilon_1)$, cannot be covered by a finite number of the sets A_λ. Again using that (X, ρ) is totally bounded it follows that $S(x_1, \varepsilon_1)$ can be covered by a finite number of spheres of radius ε_1. In a similar manner, there exists at least one of these spheres, say $S(x_2, \varepsilon_2)$, such that

$$S(x_2, \varepsilon_2) \cap S(x_1, \varepsilon_1) \neq \varnothing,$$

and which cannot be covered by a finite number of the sets A_λ.

In this way a sequence (x_n) can be defined by induction such that

$$S(x_n, \varepsilon_n) \cap S(x_{n-1}, \varepsilon_{n-1}) \neq \varnothing,$$

and for which $S(x_n, \varepsilon_n)$ cannot be covered by a finite number of the sets A_λ. (The points x_n are analogues of the mid-points of the intervals $[a_n, b_n]$ in the proof of Proposition 5.3.1.)

II. *To show that* (x_n) *is convergent.*

First it is shown that (x_n) is a fundamental sequence. Let x' be any point common to $S(x_{n+1}, \varepsilon_{n+1})$, $S(x_n, \varepsilon_n)$. Then

$$\rho(x_{n+1}, x_n) \leqslant \rho(x_{n+1}, x') + \rho(x', x_n) < \varepsilon_{n+1} + \varepsilon_n < 2\varepsilon_n;$$

therefore

$$\begin{aligned} \rho(x_{n+p}, x_n) &\leqslant \rho(x_{n+p}, x_{n+p-1}) + \rho(x_{n+p-1}, x_{n+p-2}) + \ldots + \rho(x_{n+1}, x_n) \\ &< 2\varepsilon_{n+p-1} + 2\varepsilon_{n+p-2} + \ldots + 2\varepsilon_n \\ &= 2r(2^{-(n+p-1)} + 2^{-(n+p-2)} + \ldots + 2^{-n}). \end{aligned}$$

Hence $\rho(x_{n+p}, x_n) < r.2^{-n+2}$, which holds for all $p \geqslant 0$; thus (x_n) is a fundamental sequence, so has a limit ξ (say) in X.

III. *To obtain a contradiction.*

Since $\xi \in X$, there exists λ_0 in Λ such that $\xi \in A_{\lambda_0}$; moreover, since A_{λ_0} is open there exists $\delta > 0$ such that $S(\xi, \delta) \subseteq A_{\lambda_0}$. Now take n sufficiently large, with value n_0 say, such that $\rho(x_{n_0}, \xi) < \frac{1}{2}\delta$ and $\varepsilon_{n_0} < \frac{1}{2}\delta$. Then, if $\rho(x, x_{n_0}) < \varepsilon_{n_0}$, it follows that

$$\rho(x, \xi) \leqslant \rho(x, x_{n_0}) + \rho(x_{n_0}, \xi) < \delta,$$

so that
$$S(x_{n_0}, \varepsilon_{n_0}) \subseteq S(\xi, \delta) \subseteq A_{\lambda_0}.$$

Therefore $S(x_{n_0}, \varepsilon_{n_0})$ is covered by a finite number of the sets A_λ in $\mathscr{A}$. Thus we have a contradiction, so (X, ρ) is compact.

5.4 The second equivalent characterization of compactness

We now consider the question of extending the Bolzano-Weierstrass theorem. It turns out, quite simply, that the generalization of this result holds for any metric space which is compact, but not otherwise.

THEOREM 5.4.1. *If a metric space (X, ρ) is compact, then every sequence in (X, ρ) will have at least one cluster value (in X).*

Proof. This is similar to that of Theorem 5.3.2.

Let (X, ρ) be a compact metric space, and let (x_n) be a sequence in X. Suppose, if possible, that the sequence has no cluster value (in X). Let $x \in X$, so x is not a cluster value of (x_n); then by Theorem 2.9.1 it is *not* true that, given any $\varepsilon > 0$ and any integer $m > 0$, there exists an integer $n > m$ such that $\rho(x_n, x) < \varepsilon$. Therefore, by the principles of §1.7, there exists $\varepsilon_0 > 0$ and a positive integer m_0 such that there is no $n > m_0$ for which $\rho(x_n, x) < \varepsilon_0$; hence $\rho(x_n, x) < \varepsilon_0$ for at most a finite number of values of n.

Such an open sphere $S(x, \varepsilon_0)$ can be defined for each x in X, so the set of all such spheres forms an open covering of X; since (X, ρ) is compact this must contain a finite subcovering. Each sphere of such a subcovering contains x_n for only a finite number of values of n, so their union, that is, X, contains x_n for only a finite number of values of n. This is impossible, so the theorem is established.

Next the converse result is established.

THEOREM 5.4.2. *If (X, ρ) is a metric space in which every sequence possesses at least one cluster value, then (X, ρ) is compact.*

Proof. It will be shown that (X, ρ) is complete and totally bounded so that (X, ρ) is compact.

Let (x_n) be an arbitrary fundamental sequence in (X, ρ); let c be a cluster value of this sequence. Then by Theorem 4.1.3, (x_n) converges to c; hence (X, ρ) is complete.

The proof that (X, ρ) is totally bounded will be by contradiction; thus suppose that there exists at least one $\varepsilon > 0$, say ε_0, such that there is no finite covering of X by open spheres of radius ε_0. Define a sequence (x_n) by induction as follows. Let x_1 be an arbitrary point of X. Having defined $x_1, \ldots, x_n$, let x_{n+1} be any point which is not in

$$S(x_1, \varepsilon_0) \cup \ldots \cup S(x_n, \varepsilon_0);$$

this is possible since no finite number of open spheres of radius ε_0 can cover X.

This sequence does not have a cluster value. For, clearly,

$$\rho(x_m, x_n) \geqslant \varepsilon_0 \quad \text{for all } m, n \ (m \neq n); \tag{5.4.1}$$

now if the sequence did have a cluster value, say c, then there would be a subsequence converging to c. This would be a fundamental subsequence, which contradicts (5.4.1). Therefore (x_n) has no cluster value; this gives the required contradiction.

EXERCISES 5.4

1. If (X, ρ) is compact show, by suitably modifying the proof of Theorem 5.4.1, that any infinite set in X possesses at least one limit point.

5.5 The third equivalent characterization of compactness

Finally we come to the question of the equivalence of the natural generalizations of (iii), (iv) of Proposition 5.1.4. In view of their similarity (bearing in mind the equivalent characterization for a limit point given in Theorem 2.5.1), it is reasonable to expect that it would be easier to prove that they are mutually equivalent rather than prove they are each equivalent to the generalization of (i). This is indeed the case, and therefore we establish the following result.

THEOREM 5.5.1. *Let (X, ρ) be a metric space; then the following statements are equivalent:*

(i) *every infinite set in (X, ρ) possesses at least one limit point (in X);*

(ii) *every infinite sequence in (X, ρ) possesses at least one cluster value (in X).*

Proof. Assume (i), and let (x_n) be a sequence in (X, ρ). Then, either, at least one point recurs infinitely often in (x_n), and so will be a cluster value, or no point recurs infinitely often, so (x_n) contains an infinite number of distinct points. In the latter case let Y be the infinite set consisting of these points, having a limit point c, say.

A subsequence of (x_n) which converges to c can be constructed iteratively, as follows. Having defined n_{k-1}, let n_k be the smallest integer such that $n_k > n_{k-1}$ and

$$0 < \rho(c, x_{n_k}) < k^{-1};$$

then (x_{n_k}) converges to c.

Assume (ii), and let Y be any infinite subset of X; construct any sequence (x_n) of distinct points of Y. By hypothesis there exists a convergent subsequence (lying in Y) with a limit c, say; hence by Theorem 2.5.1, c is a limit point of Y.

5.6 Summary of equivalent characterizations of compactness

We can summarize the results of the previous three sections as follows.

THEOREM 5.6.1. *Let (X, ρ) be a metric space. The following statements are equivalent:*

(i) *(X, ρ) is compact;*

(ii) *(X, ρ) is complete and totally bounded;*

(iii) *every infinite set in X possesses at least one limit point;*

(iv) *every sequence in X possesses at least one cluster value.*

Clearly (i), (iii), (iv) of the above theorem are natural generalizations of (i), (iii), (iv) of Proposition 5.1.4. Furthermore it is a simple result (which will be given in §5.7) that for R (with Euclidean metric), a set is totally bounded if and only if it is bounded; moreover, in R (which is complete) any proper subset is complete if and only if it is closed (Theorem 4.1.1). Thus (ii) of Proposition 5.1.4 could have been replaced equivalently by

(ii)′ (Y, ρ_Y) is complete and totally bounded.

With this modification, it is seen that the whole of Theorem 5.6.1 can be regarded as a natural generalization of results concerning the real line.

There are various ways in which the statements of Theorem 5.6.1 can be deduced from each other, apart from those described in §§5.3–5.5; in particular see Exercises 5.4.1, 5.6.1.

The reader is warned that, although we have the tidy situation in which all four properties (i)–(iv) above are equivalent in any metric space, this is not true in the more general situation of arbitrary topological spaces; for example, in any topological space (i) implies (iii), but not vice versa.

EXERCISES 5.6

1. Let (X, ρ) be complete and totally bounded. Prove that any sequence (x_n) in X possesses at least one cluster value by the following steps.

(i) For each k in N there exists a finite k^{-1}-net, A_k say, in X.

About each of the points of A_1 describe a sphere of radius ε_1. At least one of these spheres, say S_1, contains an infinite subsequence $(x_n^{(1)})$ of (x_n). About each of the points of A_2 describe a sphere of radius ε_2. At least one of these spheres, say S_2, contains an infinite subsequence $(x_n^{(2)})$ of $(x_n^{(1)})$. Proceed, by induction, to obtain an infinite sequence of sequences, each a subsequence of the previous one.

(ii) Consider the 'diagonal' sequence $(x_n^{(n)})$, that is, $(x_1^{(1)}, x_2^{(2)}, \ldots)$; prove that it is a subsequence of (x_n) and that it is a fundamental sequence. Deduce that (x_n) must therefore have a cluster value.

2. Show that a metric space is totally bounded if and only if every sequence in it contains a fundamental subsequence.

3. Let $\{A_\lambda : \lambda \in \Lambda\}$ be an open covering of a metric space (X, ρ). Any number $\delta > 0$ such that for each x in X there exists λ in Λ (dependent on x) for which

$$S(x, \delta) \subseteq A_\lambda$$

is called a *Lebesgue number* of the covering $\{A_\lambda : \lambda \in \Lambda\}$.

(i) Let (X, ρ) be compact and suppose, if possible, that $\{A_\lambda : \lambda \in \Lambda\}$ is an open covering of (X, ρ) not possessing a Lebesgue number. Show that there exists a convergent sequence (x_n) in X such that $S(x_n, n^{-1})$ is not contained in any of the sets A_λ.

Let x_0 denote the limit of (x_n); let $x_0 \in A_{\lambda_0}$ and $\delta_0 > 0$ be such that $S(x_0, 2\delta_0)$ is contained in A_{λ_0}. Show that there exists $n > 1/\delta_0$ for which $\rho(x_n, x_0) < \delta_0$. Hence obtain a contradiction.

Deduce that every compact metric space possesses a Lebesgue number. [An alternative proof of this result is suggested in Exercise 5.8.7.]

(ii) Give an example of a totally bounded space and an open covering of it which does not possess a Lebesgue number.

4. A metric space is said to be *countably compact* if every countable open covering contains a finite subcovering. Trivially a compact metric space is countably compact; in this exercise we outline a proof of the converse.

Let (X, ρ) be countably compact. Suppose, if possible, (X, ρ) contains an infinite sequence (x_n) of distinct points with no cluster point; let $Y = \{x_n : n \in \mathrm{N}\}$. Then for each n there exists $\varepsilon_n > 0$ such that $S_n \cap Y = \{x_n\}$ where $S_n = S(x_n, \varepsilon_n)$; for each $y \notin Y$ there exists $\varepsilon_y > 0$ such that $S_y \cap Y = \varnothing$ where $S_y = S(y, \varepsilon_y)$. Let U denote the union of all the S_y. Then $\{U, S_1, S_2, \ldots\}$ is a countable open covering of (X, ρ) which does not contain a finite subcovering.

Using also Exercise 5.2.5, deduce the following result.

For any metric space (X, ρ) the following statements are equivalent:

(i) (X, ρ) is compact;

(ii) (X, ρ) is countably compact;

(iii) every nested sequence of non-empty closed sets in (X, ρ) has non-empty intersection.

5. Show that the metric space defined in Exercise 4.3.2 is not compact by each of the following methods:

(i) find an open covering of the space which does not contain a finite subcovering;

(ii) find a sequence in the space which does not contain a convergent subsequence;

(iii) use the characterization of compactness given in (iii) of the previous exercise.

5.7 Compact subsets

So far compactness has only been discussed in relation to 'whole' metric spaces. It may happen that, although a given metric space (X, ρ) is not compact, there do exist subsets of X which possess those properties which characterize compactness; for example, bounded closed subsets of R with the Euclidean metric. This leads us to make the following definition.

DEFINITION 5.7.1. A non-empty subset Y of a metric space (X, ρ) is said to be

(i) a *compact* subset of (X, ρ) if (Y, ρ_Y) is compact;

(ii) a *totally bounded* subset of (X, ρ) if (Y, ρ_Y) is totally bounded.

The following is a straightforward result already referred to in §5.6.

LEMMA 5.7.1. *Let Y be any non-empty subset of* (R, d); *then Y is totally bounded if and only if Y is bounded.*

Proof. By Lemma 5.2.1 it follows that if Y is totally bounded then Y is bounded.

Now assume that Y is bounded so there exists $M > 0$ such that $Y \subseteq [-M, M]$. Divide $[-M, M]$ into a finite number of closed intervals, each of length less than ε. From each of those intervals which meet Y choose a point of Y. These points form a finite ε-net, so Y is totally bounded.

It is easy to see that the result of Lemma 5.7.1 extends to (R^m, d) for any $m \geqslant 1$.

A non-empty subset Y of a metric space (X, ρ) is totally bounded if and only if, given any $\varepsilon > 0$, there exists a finite set $A_\varepsilon \subseteq Y$ such that

$$\bigcup_{a \in A_\varepsilon} S_Y(a, \varepsilon) = Y, \tag{5.7.1}$$

where $S_Y(z, \varepsilon)$ is the sphere of (Y, ρ_Y) centre z and radius ε. Denoting, as usual, the sphere in (X, ρ) with centre z and radius ε by $S(z, \varepsilon)$, it is clear that (5.7.1) is equivalent to

$$\bigcup_{a \in A_\varepsilon} S(a, \varepsilon) \supseteq Y. \tag{5.7.2}$$

The next result asserts that in (5.7.2) it is not necessary for the points in A_ε to be in Y.

LEMMA 5.7.2. *Let Y be a non-empty subset of a metric space (X, ρ). Then the following statements are equivalent:*

(i) *Y is totally bounded;*

(ii) *given any $\varepsilon > 0$, there exists a finite set $B_\varepsilon \subseteq X$ such that*

$$\bigcup_{b \in B_\varepsilon} S(b, \varepsilon) \supseteq Y.$$

Proof. It is immediate that (i) implies (ii).

Now suppose that (ii) holds. Then given any $\varepsilon > 0$, let $B_{\frac{1}{2}\varepsilon}$ be a finite subset of X such that

$$\bigcup_{b \in B_{\frac{1}{2}\varepsilon}} S(b, \tfrac{1}{2}\varepsilon) \supseteq Y.$$

If $B_{\frac{1}{2}\varepsilon} \subseteq Y$, there is clearly nothing to prove.

However if $B_{\frac{1}{2}\varepsilon} \nsubseteq Y$, we construct a finite ε-net A_ε as follows. For each b in $B_{\frac{1}{2}\varepsilon}$, either $b \in Y$ or $b \notin Y$. In the former case put b in A_ε. In the latter case there are two possibilities; either $S(b, \frac{1}{2}\varepsilon)$ meets Y or it does not. In the first event take a point b' common to $S(b, \frac{1}{2}\varepsilon)$ and Y, and put it in A_ε; in the second event put no corresponding point in A_ε. The set A_ε formed in this way must be finite; it is easily seen to be an ε-net for Y. For if $y \in Y$, there exists a point b in $B_{\frac{1}{2}\varepsilon}$ such that $\rho(y, b) < \frac{1}{2}\varepsilon$; if this b is in A_ε there is nothing to prove. Otherwise b was replaced by b' where

$$\rho(y, b') \leqslant \rho(y, b) + \rho(b, b') < \varepsilon.$$

Thus A_ε is an ε-net of Y. Hence (ii) implies (i).

LEMMA 5.7.3. *Any subset of a totally bounded metric space is totally bounded.*

Proof. This result follows immediately from Lemma 5.7.2.

The question which now naturally arises is 'when is a non-empty subset Y of a compact metric space (X, ρ) compact?' Since a metric space is compact if and only if it is complete and totally bounded, it follows from Lemma 5.7.3 that the answer to our question is 'when and only when (Y, ρ_Y) is complete'. Since (X, ρ) is complete, by Theorem 4.1.1, (Y, ρ_Y) is complete if and only if Y is closed in (X, ρ). Hence the following result is established.

LEMMA 5.7.4. *If Y is a non-empty subset of a compact metric space (X, ρ), then (Y, ρ_Y) is compact if and only if Y is closed in (X, ρ).*

In view of Theorem 4.1.2 we also have the following result.

LEMMA 5.7.5. *If Y is a non-empty subset of a metric space (X, ρ) and if (Y, ρ_Y) is compact, then Y is closed in (X, ρ).*

Since compactness is such an important concept, it is sometimes desirable to be able to consider sets which are not closed, so will not be compact, but which have compact closures. This leads to

DEFINITION 5.7.2. A non-empty subset Y of a metric space (X, ρ) is said to be a *relatively compact* subset of (X, ρ) if $\bar{Y}$ is a compact subset of (X, ρ).

The following result is a variation of Theorem 5.6.1.

LEMMA 5.7.6. *Let Y be a non-empty subset of a metric space (X, ρ). Then the following statements are equivalent:*

(i) *Y is relatively compact;*

(ii) *every infinite set in Y possesses at least one limit point (not necessarily in Y);*

(iii) *every sequence in Y possesses at least one cluster point (not necessarily in Y).*

Proof. Assume (i); then $\bar{Y}$ is compact so (ii) follows immediately from Theorem 5.6.1.

Assume (ii); then we can deduce (iii) by exactly the same steps as used in the proof of Theorem 5.5.1.

Assume (iii). Let (x_n) be a sequence in $\bar{Y}$; we first replace this by a sequence (x_n') in Y such that $\rho(x_n, x_n') \to 0$ as $n \to \infty$. Since $x_n \in \bar{Y}$ there exists x_n' in Y such that $\rho(x_n, x_n') < 1/n$; thus (x_n') is a sequence in Y, so there exists a convergent subsequence (x_{nk}') which converges to some element c, which must be in $\bar{Y}$. Now

$$\rho(x_{nk}, c) \leqslant \rho(x_{nk}, x_{nk}') + \rho(x_{nk}', c);$$

letting $k \to \infty$, $x_{nk} \to c$, so c is a cluster point of (x_n). Hence by Theorem 5.6.1, $\bar{Y}$ is compact, that is, Y is relatively compact. This establishes (i).

EXERCISES 5.7

1. Which of the sets

$$[0, 1], (0, 1], \{1, 2, 3, 4\}, \mathsf{Z}, \mathsf{Q} \cap [0, 1], \mathsf{Q} \cap (0, \sqrt{2})$$

are compact in (R, d)? For those which are not compact find an open covering which does not contain a finite subcovering.

2. Let Y be a finite subset of a metric space (X, ρ); show that Y is compact.

If ρ' is any discrete metric associated with X, show that (X, ρ') contains no infinite subset which is compact.

3. (i) If Y_1, Y_2 are compact subsets of a metric space (X, ρ), show that $Y_1 \cap Y_2$, $Y_1 \cup Y_2$ are also compact.

(ii) If $\{Y_\lambda : \lambda \in \Lambda\}$ is any collection of compact subsets of (X, ρ), and if

$$V = \bigcap_{\lambda \in \Lambda} Y_\lambda, \quad W = \bigcup_{\lambda \in \Lambda} Y_\lambda,$$

prove that V is compact, but that W is not necessarily compact.

4. If Y_1, Y_2 are subsets of a metric space (X, ρ) such that Y_1 is closed and Y_2 is compact, show that $Y_1 \cap Y_2$ is compact.

5. Let Y be a subset of a metric space (X, ρ) and let Y' denote the set of limit points of Y.

(i) Show that Y is totally bounded if and only if $\bar{Y}$ is totally bounded.

(ii) If Y is totally bounded show that Y' is totally bounded.

(iii) If Y is compact show that Y' is compact.

(iv) Give an example in which Y is totally bounded and Y' is not compact.

6. Prove that Y is a compact subset of a metric space (X, ρ) if and only if every covering of Y by open subsets of (X, ρ) contains a finite subcovering.

7. Let f be a mapping of a metric space (X, ρ) into itself such that $\rho(f(x), f(y)) < \rho(x, y)$ for all distinct x, y in X.

(i) Show that f cannot have more than one fixed point.

(ii) Show that f is continuous over X.

(iii) Let x_0 be an arbitrary point of X and let $x_n = f^{(n)}(x_0)$. Suppose that (x_n) contains a convergent subsequence (x_{n_k}) with limit y. Show that for any positive integer p

$$f^{(p)}(y) = \lim_{k \to \infty} x_{n_k + p},$$

and that $(\rho(x_n, x_{n+1}))$ is a decreasing sequence. Deduce that

$$\rho(y, f(y)) = \rho(f(y), f^{(2)}(y)).$$

Hence show that y is a fixed point of f.

(iv) If $f(X)$ is a compact subset of (X, ρ), deduce that f has a unique fixed point.

5.8 Properties of functions continuous over compact sets

In this section some fundamental theorems concerning real-valued functions which are continuous over closed bounded intervals of R (with Euclidean metric) will be generalized, in a natural way, to results concerning functions which are continuous over compact subsets of arbitrary metric spaces.

The first result to be generalized is

PROPOSITION 5.8.1. *Let $I = [a, b]$; if the function $f: I \to \mathsf{R}$ is continuous over I then its range $f(I)$ is bounded. Moreover, the supremum and infimum of $f(I)$ are attained; that is, if $M = \sup f(I)$, $m = \inf f(I)$, then there exist α, β in I such that $f(\alpha) = M, f(\beta) = m$.*

The generalization of the first part of this result is as follows.

THEOREM 5.8.1. *Let (X, ρ), (X', ρ') be two metric spaces, and Y be a compact subset of (X, ρ); if $f:(Y, \rho_Y) \to (X', \rho')$ is continuous then $f(Y)$ is a compact subset of (X', ρ').*

Proof. For brevity let $Y' = f(Y)$, so it is necessary to show that, if f is a continuous mapping of (Y, ρ_Y) onto $(Y', \rho'_{Y'})$, then compactness of the first metric space implies compactness of the second; for further brevity the subscripts of ρ_Y, $\rho'_{Y'}$ are dropped.

Let $\{Z_\lambda : \lambda \in \Lambda\}$ be any open covering of (Y', ρ'). Since f is continuous, $f^{-1}(Z_\lambda)$ is open in (Y, ρ). Furthermore,

$$\{f^{-1}(Z_\lambda) : \lambda \in \Lambda\}$$

is an open covering of (Y, ρ); why? Since Y is compact, there exists a finite subset Λ' of Λ such that

$$\{f^{-1}(Z_\lambda) : \lambda \in \Lambda'\}$$

also covers (Y, ρ); it is finally shown that $\{Z_\lambda : \lambda \in \Lambda'\}$ is a finite subcovering of (Y', ρ') so that the latter is compact. Now

$$\begin{aligned} Y' = f(Y) &= f\Big(\bigcup_{\lambda \in \Lambda'} f^{-1}(Z_\lambda)\Big) \\ &= \bigcup_{\lambda \in \Lambda'} f(f^{-1}(Z_\lambda)) \subseteq \bigcup_{\lambda \in \Lambda'} Z_\lambda \end{aligned}$$

using (1.2.6), (1.2.14). However $Z_\lambda \subseteq Y'$ for all λ in Λ; so that

$$Y' = \bigcup_{\lambda \in \Lambda'} Z_\lambda.$$

This completes the proof.

Theorem 5.8.1 can be summarized as 'a continuous image of a compact set is compact'. An alternative proof of this result is contained in Exercise 5.8.1.

Next the second part of Proposition 5.8.1 is generalized. First a simple result is established.

LEMMA 5.8.1. *Let A be a closed, bounded (that is compact) subset of* (R, d). *Let $M = \sup A$, $m = \inf A$; then $M, m \in A$.*

Proof. For any $\varepsilon > 0$ there exists a in A such that $M - \varepsilon < a \leqslant M$; hence M is a limit point of A, and so $M \in A$. Similarly $m \in A$.

THEOREM 5.8.2. *Let Y be a compact subset of a metric space (X, ρ). If $f:(Y, \rho_Y) \to (\mathsf{R}, \mathsf{d})$ is continuous, then the supremum and infimum of $f(Y)$ are attained; that is, if $M = \sup f(Y)$, $m = \inf f(Y)$, there exist α, β in Y such that $f(\alpha) = M$, $f(\beta) = m$.*

Proof. Since $f(Y)$ is a compact subset of (R, d) it is closed and bounded. Hence $f(Y)$ possesses a supremum M and an infimum m. By Lemma 5.8.1, $M, m \in f(Y)$, so there exist α, β in Y such that $f(\alpha) = M$, $f(\beta) = m$.

In §3.3 it was explained that although a bijective mapping must have an inverse, the inverse of a continuous bijective function is not necessarily continuous. However, we do have the following result which is deduced from Theorem 5.8.1.

THEOREM 5.8.3. *If (X, ρ) is compact and $f:(X, \rho) \to (X', \rho')$ is a continuous bijective function, then the inverse $f^{-1}:(X', \rho') \to (X, \rho)$ is also continuous, (that is, f is a homeomorphism).*

Proof. Let Y be a closed subset of (X, ρ); then by Lemma 5.7.4, Y is a compact subset of (X, ρ). Therefore, by Theorem 5.8.1, $f(Y)$ is a compact, and hence closed, subset of (X', ρ'). But $f(Y) = (f^{-1})^{-1}(Y)$ so $(f^{-1})^{-1}(Y)$ is a closed subset of (X', ρ'). Hence, by Theorem 3.2.1, f^{-1} is continuous.

The second result to be generalized is

Proposition 5.8.2. *Let $I = [a, b]$; if the function $f: I \to \mathsf{R}$ is continuous over I, then it is uniformly continuous there.*

The reader is reminded that uniform continuity was defined for arbitrary metric spaces in §3.1. The generalization of Proposition 5.8.2 that we shall prove is as follows.

Theorem 5.8.4. *Let (X, ρ), (X', ρ') be two metric spaces, and Y be a compact subset of (X, ρ); if $f:(Y, \rho_Y) \to (X', \rho')$ is continuous over Y, then it is uniformly continuous there.*

Proof. Given any $\varepsilon > 0$, for each point y of Y, there exists $\delta_y > 0$ for which $\rho'(f(y), f(y')) < \frac{1}{2}\varepsilon$ for all y' in Y such that $\rho(y, y') < \delta_y$. Since $\{S_Y(y, \frac{1}{2}\delta_y): y \in Y\}$ is an open covering of Y, there exists a finite subcovering of Y by, say,

$$S_Y(y_1, \tfrac{1}{2}\delta_1), \ldots, S_Y(y_n, \tfrac{1}{2}\delta_n) \tag{5.8.1}$$

where we write δ_i for δ_{y_i} $(i = 1, \ldots, n)$. Let $\delta^* = \min(\frac{1}{2}\delta_1, \ldots, \frac{1}{2}\delta_n)$.

Now let y, y' be any pair of elements of Y such that $\rho(y, y') < \delta^*$; since $y \in Y$, it must belong to one of the spheres of (5.8.1), say $S_Y(y_k, \frac{1}{2}\delta_k)$. Therefore $\rho(y, y_k) < \frac{1}{2}\delta_k$, and hence

$$\rho(y', y_k) \leqslant \rho(y', y) + \rho(y, y_k) < \delta_k.$$

Hence, for such y, y'

$$\rho'(f(y), f(y')) \leqslant \rho'(f(y), f(y_k)) + \rho'(f(y_k), f(y')) < \tfrac{1}{2}\varepsilon + \tfrac{1}{2}\varepsilon = \varepsilon.$$

This completes the proof since δ^* is independent of the points y, y'.

We conclude this section with two remarks.

(i) It is largely the properties possessed by functions which are continuous over compact sets, as described in Theorems 5.8.1, 5.8.2, 5.8.3, 5.8.4, that make compactness an important concept.

(ii) Observe that we have frequently studied (in particular in §§2.2, 2.4, 4.2) the space $(\mathscr{C}(I), \sigma)$ where $I = [a, b]$ and σ denotes the supremum metric. If $\mathscr{C}(X)$ denotes the set of all real-valued continuous functions defined on a compact space (X, ρ) then the above results enable us to generalize our study of $(\mathscr{C}(I), \sigma)$ to that of $(\mathscr{C}(X), \sigma)$ where σ is again the supremum metric

$$\sigma(f, g) = \sup_{x \in X} |f(x) - g(x)|.$$

Moreover $(\mathscr{C}(X), \sigma)$ is complete. The verification that σ is a metric and that the space is complete is a straightforward extension of the argument used when $X = I$; it is necessary to use the result of Exercise 3.1.2 instead of Theorem 1.6.7. The details are left to the reader. The space $(\mathscr{C}(X), \sigma)$ is of considerable importance in more abstract analysis; however we shall refer to it again only in §7.8.

Exercises 5.8

1. Give an alternative proof of Theorem 5.8.1 using the equivalent characterization of compactness established in §5.4. [However such a proof does not generalize to arbitrary topological spaces as does the proof given in §5.8.]

2. Give examples to show that the criterion in Theorems 5.8.1, 5.8.2, 5.8.3, 5.8.4 that Y be a compact subset of (X, ρ) cannot be dropped (for the case of Theorem 5.8.3 consider the function defined in Exercise 3.3.6). [See also Exercise 7.3.3.]

3. Let f be a mapping of a compact metric space (X, ρ) into itself such that $f(X)$ is everywhere dense in (X, ρ), and

$$\rho(f(x), f(y)) = \rho(x, y)$$

for all x, y in X. Prove that $f(X) = X$ (so that f is an isometry).

4. Let (X, ρ) be totally bounded; if $f:(X, \rho) \to (X', \rho')$ is uniformly continuous, prove that $f(X)$ is a totally bounded subset of (X', ρ'). [Thus if f is a bijection from (X, ρ) to (X', ρ') such that f and f^{-1} are both uniformly continuous, then (X, ρ) is totally bounded if and only if (X', ρ') is totally bounded.]

5. Let Y, Z be non-empty subsets of a metric space (X, ρ). Prove that

(i) if Y is compact then there exists y in Y such that $\rho(y, Z) = \rho(Y, Z)$;

(ii) if Y and Z are both compact, then there exists y in Y and z in Z such that $\rho(y, z) = \rho(Y, Z)$;

(iii) if Y is compact and Z is closed, then $\rho(Y, Z) = 0$ if and only if $Y \cap Z \neq \varnothing$.

6. Let f be a mapping of a metric space (X, ρ) into itself such that $\rho(f(x), f(y)) < \rho(x, y)$ for all distinct x, y in X. Suppose that (X, ρ) is compact. Let

$$h = \inf_{x \in X} \rho(x, f(x));$$

by using a contradiction argument show that $h = 0$.

Using (i) of Exercise 5.7.7 deduce that f possesses a unique fixed point. [This exercise provides a simpler proof of a weaker form of the result of Exercise 5.7.7.]

7. By imitating the first part of the proof of Theorem 5.8.4 obtain an alternative proof of the result (i) of Exercise 5.6.3.

5.9 Examples concerning compactness in metric spaces

In this, and the following, section we shall look at some of the metric spaces defined in §2.2 and seek necessary and sufficient conditions for subsets of them to be compact. As before, the Roman numerals below correspond to those of §2.2.

(i), (ii) $X = \mathsf{R}, \mathsf{R}^m$ with the corresponding Euclidean metric.

The situation in these cases is well known. The metric spaces are not compact; for example in (R, d) the open covering $\{(-n, n) : n \in \mathsf{N}\}$ contains no finite subcovering. A subset is compact if and only if it is closed and bounded.

(iii), (iv) $X = \mathsf{R}^m$, and let

$$\rho(x, y) = \{\textstyle\sum_i |x_i - y_i|^p\}^{1/p}$$

where $p \geqslant 1$, and

$$\rho'(x, y) = \max_i |x_i - y_i|.$$

Then, as was seen in §3.4, the metrics ρ, ρ' are equivalent; hence, if $Y \subseteq \mathsf{R}^m$, the spaces (Y, ρ_Y), (Y, ρ'_Y) are homeomorphic, and by Theorem 5.2.1, either of these metric spaces is compact if and only if the other is compact. When $p = 2$, ρ is the Euclidean metric; thus (R^m, ρ), (R^m, ρ') are not compact, and a subset of R^m is compact in (R^m, ρ) or (R^m, ρ') if and only if it is closed and bounded.

(v) X is any non-empty set and ρ is the standard discrete metric. Any subset Y of (X, ρ) is compact if and only if it is finite; this is seen as follows.

Suppose that Y is finite, and $\{Y_\lambda : \lambda \in \Lambda\}$ is an open covering of (Y, ρ_Y); let $Y = \{y_1, \ldots, y_n\}$. Then there exist $Y_{\lambda_i}, i = 1, \ldots, n$ such that $y_i \in Y_{\lambda_i}$; hence $\{Y_{\lambda_i} : i = 1, \ldots, n\}$ is a finite subcovering of (Y, ρ_Y), so Y is compact.

Conversely suppose Y is a compact subset of (X, ρ). Since every subset of X is open in (X, ρ), $\{\{y\} : y \in Y\}$ is an open covering of Y. This must contain a finite subcovering of Y, that is, there exist $y_1, \ldots, y_n$ in Y such that

$$\{y_1\} \cup \ldots \cup \{y_n\} = Y,$$

so Y is finite.

In view of (i) of Exercise 3.4.6, a subset of any discrete space is compact if and only if it is finite.

(vi) $X = \ell^p$ and

$$\rho(x, y) = \{\textstyle\sum_i |x_i - y_i|^p\}^{1/p}$$

where $x = (x_i)$, $y = (y_i)$ are two elements of ℓ^p. The example of §5.1 shows that (ℓ^p, ρ) is not compact.

Let $Y \subset \ell^p$; then Y is a compact subset of (ℓ^p, ρ) if and only if it is closed and both of the following conditions are satisfied:

(*a*) Y is bounded, that is, there exists $M > 0$ such that

$$\left\{\sum_{i=1}^{\infty} |y_i|^p\right\}^{1/p} \leqslant M$$

for all y $(=(y_i))$ in Y, and

(*b*) given any $\varepsilon > 0$ there exists I such that

$$\left\{\sum_{i=I}^{\infty} |y_i|^p\right\}^{1/p} < \varepsilon$$

where I is independent of y $(=(y_i))$ in Y.

The proof of this result is outlined in Exercise 5.9.3.

Exercises 5.9

1. Show that the metric spaces (X, ρ), (X, ρ') defined in Exercise 2.2.2 are totally bounded (or compact) if and only if the spaces (X_i, ρ_i), $i = 1, \ldots, m$ are all totally bounded (or compact, respectively). (For the 'only if' part use Exercise 3.2.8.)

2. Let (X, ρ) be the metric space defined in Exercise 2.2.4.

(i) By considering the sequence $(x^{(n)})$ defined by $x^{(n)} = (n, n, \ldots)$, show that (X, ρ) is not compact.

(ii) Let (a_i) be such that $a_i \geqslant 0$ for all i, and let $Y\,(\subset X)$ be a closed set such that for all $y\,(=(y_i))$ in Y, $|y_i| \leqslant a_i\ (i \in \mathsf{N})$.

Let $(x^{(n)})$ be a sequence in Y where $x^{(n)} = (x_1^{(n)}, x_2^{(n)}, \ldots)$.

Show that the sequence $(x_1^{(n)})$ in (R, d) must have a convergent subsequence, limit $x_1^{(0)}$ say, where $|x_1^{(0)}| \leqslant a_1$. Consider next the subsequence of $(x^{(n)})$ whose first coordinates are precisely the subsequence of $(x_1^{(n)})$ just described. Show that the sequence of the second coordinates of the subsequence contain a subsequence convergent in (R, d), limit $x_2^{(0)}$ say, where $|x_2^{(0)}| \leqslant a_2$.

Proceeding in this way define an element $x^{(0)}$ in Y; show that $x^{(0)}$ is a cluster value of $(x^{(n)})$. Deduce that Y is a compact subset of (X, ρ).

Also extract from the above argument a convergent subsequence of $(x^{(n)})$. (Hint: see Exercise 5.6.1.)

(iii) Using the last part of Exercise 3.2.8 prove that the criterion of (ii) is not only sufficient for Y to be compact but that it is also necessary.

3. Establish the assertion of §5.9(vi) by the following steps.

First assume that Y is a compact subset of (ℓ^p, ρ). Then Y must be closed.

Y is bounded; deduce (a).

Given any $\varepsilon > 0$, let $\{y^{(1)}, \ldots, y^{(m)}\}$ be m points of Y which form a $\frac{1}{2}\varepsilon$-net of Y; set $y^{(n)} = (y_1^{(n)}, y_2^{(n)}, \ldots)$ for $n = 1, \ldots, m$. There exists I such that

$$\left\{\sum_{i=I}^{\infty} |y_i^{(n)}|^p\right\}^{1/p} < \tfrac{1}{2}\varepsilon,$$

for $n = 1, \ldots, m$; for any y in Y show that

$$\left\{\sum_{i=I}^{\infty} |y_i - y_i^{(n)}|^p\right\}^{1/p} < \tfrac{1}{2}\varepsilon$$

for some n such that $1 \leqslant n \leqslant m$. Deduce (b).

Next assume that Y is a closed set which satisfies conditions (a), (b).

Given any $\varepsilon > 0$ there exists I such that

$$\sum_{i=I+1}^{\infty} |y_i|^p < \tfrac{1}{2}\varepsilon^p$$

for all y $(=(y_i))$ in Y; associate with each y in Y the point

$$(y_1, \ldots, y_I, 0, 0, \ldots),$$

and call the set of all such points Y_ε. Let $\tilde{Y}_\varepsilon$ be the subset of R^I defined by

$$\tilde{Y}_\varepsilon = \{(y_1, \ldots, y_I) : y \in Y\}.$$

Show that $\tilde{Y}_\varepsilon$ is a totally bounded subset of (R^I, ρ^I) where

$$\rho^I(u, v) = \left\{\sum_{i=1}^{I} |u_i - v_i|^p\right\}^{1/p}$$

$u = (u_1, \ldots, u_I)$, $v = (v_1, \ldots, v_I)$, and that $(Y_\varepsilon, \rho_{Y_\varepsilon})$ is homeomorphic to the subset $\tilde{Y}_\varepsilon$ of (R^I, ρ^I). Hence obtain a finite ε-net of Y.

Deduce that Y is compact.

4. Show that a closed, bounded, subset Y of the metric space defined in Exercise 2.1.6 is compact if and only if

(i) for each $\varepsilon > 0$ the set

$$\{x_1 : \text{there exists } (x_1, x_2) \text{ in } Y \text{ such that } |x_2| \geqslant \varepsilon\}$$

is finite, and

(ii) the sets

$$Y \cap \{(x_1, x_2) : x_1 \text{ fixed}, x_2 \in \mathsf{R}\}, \quad Y \cap \{(x_1, 0) : x_1 \in \mathsf{R}\}$$

are compact in the Euclidean sense.

5. Find necessary and sufficient conditions for a subset of the metric space defined in Exercise 2.1.5 to be compact.

5.10 A further example: the Arzelà-Ascoli theorem

Throughout this section, I denotes the closed interval $[a, b]$ and ρ is the usual supremum metric on $\mathscr{C}(I)$. Since the study of continuous functions is a foremost part of analysis it is useful to know which subsets of $(\mathscr{C}(I), \rho)$ are compact. In this section, after the definition of two simple terms, necessary and sufficient conditions for this will be obtained.

DEFINITION 5.10.1. A set A of real-valued functions f, each of which is defined on a set S, is said to be *uniformly bounded over S* if there exists a constant M such that $|f(x)| < M$ for all f in A and all x in S.

LEMMA 5.10.1. *Let $Y \subseteq \mathscr{C}(I)$; then Y is uniformly bounded if and only if Y is a bounded subset of $(\mathscr{C}(I), \rho)$.*

Proof. This is left to the reader.

DEFINITION 5.10.2. A set A of real-valued functions f, each of which is defined on a subset S of a metric space (X, ρ), is said to be *equicontinuous over* S if, given any $\varepsilon > 0$, there exists $\delta > 0$ such that $|f(x_1)-f(x_2)| < \varepsilon$ for all f in A and all x_1, x_2 in S for which $\rho(x_1, x_2) < \delta$.

Note that this immediately implies that each individual function is uniformly continuous over S.

THEOREM 5.10.1 (The Arzelà-Ascoli theorem). *Let A be a subset of $(\mathscr{C}(I), \rho)$. Then the following statements are equivalent:*

(i) *A is compact;*

(ii) *A is a closed subset of $(\mathscr{C}(I), \rho)$ and is both uniformly bounded and equicontinuous over I.*

Proof. First assume (i). Then (A, ρ_A) is compact, so A is closed in $(\mathscr{C}(I), \rho)$. Since A is totally bounded, it is a bounded subset of $(\mathscr{C}(I), \rho)$ and so, by Lemma 5.10.1, A is uniformly bounded over I.

For any $\varepsilon > 0$, there exists a finite $\frac{1}{3}\varepsilon$-net of A, so there is a finite set $\{f_1, \dots, f_k\}$ of functions in A such that, for any f in A, there exists i $(1 \leqslant i \leqslant k)$ for which $\rho(f, f_i) < \frac{1}{3}\varepsilon$, that is, for which

$$\sup_t |f(t)-f_i(t)| < \tfrac{1}{3}\varepsilon.$$

Each f_i is continuous over I so is uniformly continuous there; hence there exists $\delta_i > 0$ such that

$$|f_i(t_1)-f_i(t_2)| < \tfrac{1}{3}\varepsilon$$

for all t_1, t_2 for which $|t_1-t_2| < \delta_i$. Let $\delta = \min(\delta_1, \dots, \delta_k)$. Then for $|t_1-t_2| < \delta$ and any f in A,

$$|f(t_1)-f(t_2)| \leqslant |f(t_1)-f_i(t_1)|+|f_i(t_1)-f_i(t_2)|+|f_i(t_2)-f(t_2)| < \varepsilon,$$

so A is equicontinuous over I.

Now assume (ii). Since $(\mathscr{C}(I), \rho)$ is complete and A is closed, it follows that (A, ρ_A) is complete. It remains to show that the subspace is totally bounded.

Since A is uniformly bounded, there exists M such that $|f| < M$

for all f in A. Let $\varepsilon > 0$; since A is equicontinuous, there exists $\delta > 0$ for which

$$|f(t_1)-f(t_2)| < \tfrac{1}{4}\varepsilon \tag{5.10.1}$$

for all t_1, t_2 in I such that $|t_1-t_2| < \delta$, and for all f in A. Construct a partition P of I, say

$$a = x_0 < x_1 < \dots < x_n = b$$

such that $\|P\| < \delta$ (where $\|P\|$ denotes the length of the longest subinterval of P); likewise construct a partition P' of $[-M, M]$,

$$-M = y_0 < y_1 < \dots < y_m = M,$$

such that $\|P'\| < \frac{1}{4}\varepsilon$. Thus the rectangle $[a, b] \times [-M, M]$ has been divided into mn sub-rectangles.

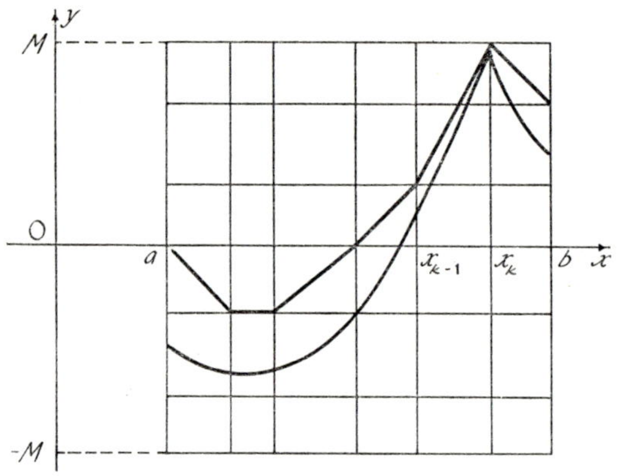

Corresponding to each function f of A, define a function $\bar{f}$ as follows. Let $\bar{f}(x_k)$ be the least y_i in P' such that $y_i \geqslant f(x_k)$, $k = 0, 1, \dots, n$; let $\bar{f}(x)$ be defined linearly between these points. In this way a unique function $\bar{f}$ (which is in $\mathscr{C}(I)$, although not necessarily in A) is associated with each of the members of A. Let $B_\varepsilon = \{\bar{f} : f \in A\}$.

Since $\bar{f}(x_k) \geqslant f(x_k)$, it follows that

$$\begin{aligned}\bar{f}(x_{k+1})-\bar{f}(x_k) &\leqslant \bar{f}(x_{k+1})-f(x_k)\\ &= \bar{f}(x_{k+1})-f(x_{k+1})+f(x_{k+1})-f(x_k) < \tfrac{1}{2}\varepsilon.\end{aligned}$$

Similarly

$$\bar{f}(x_{k+1})-\bar{f}(x_k) \geqslant f(x_{k+1})-\bar{f}(x_k)$$
$$= f(x_{k+1})-f(x_k)+f(x_k)-\bar{f}(x_k) > -\tfrac{1}{2}\varepsilon,$$

and hence

$$\left|\bar{f}(x_{k+1})-\bar{f}(x_k)\right| < \tfrac{1}{2}\varepsilon. \tag{5.10.2}$$

Then, if $x_k \leqslant t < x_{k+1}$,

$$\left|f(t)-\bar{f}(t)\right| \leqslant \left|f(t)-f(x_k)\right|+\left|f(x_k)-\bar{f}(x_k)\right|+\left|\bar{f}(x_k)-\bar{f}(t)\right|$$
$$< \tfrac{1}{4}\varepsilon+\tfrac{1}{4}\varepsilon+\tfrac{1}{2}\varepsilon$$

by (5.10.1) and (5.10.2), and since $\bar{f}$ is linear on $[-x_k, x_{k+1}]$; thus $\rho(f,\bar{f}) < \varepsilon$. We have therefore shown that given any f in A, there exists $\bar{f}$ in B_ε such that $\rho(f,\bar{f}) < \varepsilon$.

Since there are exactly $(m+1)^{n+1}$ polygonal arcs through the points (x_i, y_j), $i = 0, 1, \ldots, n$; $j = 0, 1, \ldots, m$, the ε-net just constructed must be finite. Hence, by Lemma 5.7.2, A is totally bounded.

By Theorem 5.6.1, any infinite sequence in a compact subset of $(\mathscr{C}(I), \rho)$ contains a subsequence which converges in $(\mathscr{C}(I), \rho)$; moreover, since convergence in this metric space is equivalent to uniform convergence in the Euclidean sense (see §2.4(viii)), we can restate the Arzelà-Ascoli theorem in the following form.

Corollary 5.10.1. *Any infinite sequence in a closed, bounded and equicontinuous subset A of $(\mathscr{C}(I), \rho)$ contains a subsequence which is uniformly convergent* (*in the Euclidean sense*); *moreover, the limit of this sequence belongs to A.*

In view of Lemma 5.7.6 and Exercise 5.10.1 this can be modified as follows.

Corollary 5.10.2. *Any infinite sequence in a bounded equicontinuous subset A of $(\mathscr{C}(I), \rho)$ contains a subsequence which is uniformly convergent* (*in the Euclidean sense*); *however the limit of this sequence does not necessarily belong to A.*

Exercises 5.10

1. Let Y be a subset of $(\mathscr{C}(I), \rho)$. If Y is uniformly bounded (or equicontinuous) over I, show that $\bar{Y}$ is also uniformly bounded (or equicontinuous, respectively) over I.

2. (i) Let (f_n) be a sequence of differentiable functions in $\mathscr{C}(I)$ such that

(*a*) $\sup_x |f_n'(x)| \leqslant M$ for all n, and

(*b*) there exists c in I such that the set $\{f_n(c) : n \in \mathsf{N}\}$ is bounded.

Show that (f_n) contains a subsequence which converges in $(\mathscr{C}(I), \rho)$.

(ii) Let $c \in I$, and Y be an equicontinuous set of functions $f: I \to \mathsf{R}$ such that $\{f(c) : f \in Y\}$ is a bounded subset of R. Show that Y is uniformly bounded.

Give an example of an equicontinuous set of functions which is not uniformly bounded.

3. (i) For $0 < \varepsilon < 1$ and $0 \leqslant x \leqslant 1$ show that $|1 - x^n| < \varepsilon$ if and only if $x > (1-\varepsilon)^{1/n}$. Let $I_0 = [0, 1]$. Using the result that $(1-\varepsilon)^{1/n} \to 1$ as $n \to \infty$, deduce that the sequence of functions (x^n) [sic] is not equicontinuous over I_0, and so does not possess a cluster value. Hence show that the 'unit' sphere

$$\{f\colon \sup_{x \in I_0} |f(x)| \leqslant 1\}$$

is not a compact subset of $(\mathscr{C}(I_0), \rho)$ where ρ is the usual supremum metric.

(ii) By showing that the sequence of functions (x^n) is not uniformly convergent over I_0, give an alternative proof that the sequence (x^n) does not possess a cluster value in $(\mathscr{C}(I_0), \rho)$.

4. Fill in the details of the following outline of an alternative proof of the sufficiency part of Theorem 5.10.1.

Let (f_n) be any sequence in a subset Y of $(\mathscr{C}(I), \rho)$ where Y is closed, uniformly bounded and equicontinuous over I.

(i) Express the set $H = \mathsf{Q} \cap I$ as $\{r_m : m \in \mathsf{N}\}$.

The real sequence $(f_n(r_1))$ contains a convergent subsequence; let it be denoted by $(f_{1n}(r_1))$. The real sequence $(f_{1n}(r_2))$ contains a convergent subsequence; let it be denoted by $(f_{2n}(r_2))$. Thus the sequence (f_{2n}) converges at r_1, r_2. Proceeding in this way define, for every m in N, a sequence $(f_{mn}(r_m))$ such that the sequence (f_{mn}) is convergent at $r_1, \ldots, r_m$.

(ii) The diagonal subsequence (f_{nn}) of (f_n) is convergent at all points of H.

(iii) For brevity set $g_n = f_{nn}$; given any $\varepsilon > 0$ there exists $\delta > 0$ for which $|g_n(x) - g_n(y)| < \frac{1}{3}\varepsilon$ for all n and all x, y in I such that

$|x-y| < \delta$, where δ is independent of n. Let $q_1, \ldots, q_t$ be rationals such that

$$a < q_1 < q_2 < \ldots < q_t < b$$

and
$$\max(q_1-a, q_2-q_1, \ldots, b-q_t) < \delta.$$

For $i = 1, \ldots, t$ there exists N_i such that $|g_n(q_i)-g_m(q_i)| < \frac{1}{3}\varepsilon$ for all $m, n > N_i$. Set $N = \max(N_1, \ldots, N_t)$; then $|g_m(x)-g_n(x)| < \varepsilon$ for all x in I and all $m, n > N$.

(iv) Thus Y is a compact subset of $(\mathscr{C}(I), \rho)$.

5. Let (X, ρ) be a compact metric space and let $(\mathscr{C}(X), \sigma)$ be the metric space defined at the end of §5.8. Prove that a subset of $(\mathscr{C}(X), \sigma)$ is compact if and only if it is closed and is both uniformly bounded and equicontinuous over X.

5.11 Peano's theorem

The Arzelà-Ascoli theorem is now used to obtain an important existence theorem due to Peano concerning the differential equation $y' = f(x, y)$. The statement of this result is similar to that of Theorem 4.8.1; the only difference is that we shall no longer assume that the function f satisfies a Lipschitz condition, although we shall require that f is bounded. In consequence it is no longer possible to conclude that the solution is unique; an example will be given in which there exists more than one solution.

Classically, Peano's theorem was proved by a method which involved approximate polygonal solutions (that is, functions whose graphs are polygonal arcs approximating to the exact solution); this method, although direct, is comparatively complicated. The proof given below is a less obvious, but more elegant, one due to Tonelli.

THEOREM 5.11.1. *Let* $I = [x_0-a, x_0+a]$ *and* $S = I \times \mathsf{R}$; *let* $f: S \to \mathsf{R}$ *be continuous and bounded over* S.

Then the differential equation

$$y' = f(x, y) \tag{5.11.1}$$

possesses a solution through (x_0, y_0) *over* I.

Proof. By Lemma 4.8.1 the differential equation (5.11.1), subject to the condition $y(x_0)=y_0$, is equivalent to the integral equation

$$y(x) = y_0 + \int_{x_0}^{x} f(t, y(t))\,dt \tag{5.11.2}$$

for $x \in I$. We consider first the half-interval $I_0 = [x_0, x_0+a]$.

Define a sequence (y_n) of real-valued continuous functions over I_0 as follows. For $x_0 \leqslant x \leqslant x_0 + a/n$ set $y_n(x) = y_0$; for $x_0 + a/n < x \leqslant x_0 + a$ set

$$y_n(x) = y_0 + \int_{x_0}^{x-a/n} f(t, y_n(t))\,dt. \qquad (5.11.3)$$

Note that this is a meaningful definition for $y_n(x)$; for, setting

$$J_r = [x_0 + (r-1)a/n,\ x_0 + ra/n], \quad r = 1, \ldots, n,$$

if $x \in J_r$, then $y_n(x)$ is defined in terms of $y_n(x)$ for x in $J_1 \cup \ldots \cup J_{r-1}$.

Let $|f(x, y)| \leqslant M$ for all (x, y) in S. If $x > x_0 + a/n$, then

$$|y_n(x) - y_0| \leqslant \int_{x_0}^{x-a/n} |f(t, y_n(t))|\,dt \leqslant M\,|x - an^{-1} - x_0\,| < Ma;$$

if $x_0 \leqslant x \leqslant x_0 + a/n$ then $|y_n(x) - y_0| = 0$, so the sequence (y_n) is uniformly bounded over I_0. Likewise

$$|y_n(x) - y_n(x')| \leqslant M|x - x'|$$

for all x, x' in I_0, so the sequence (y_n) is equicontinuous over I_0. Therefore by the Arzelà-Ascoli theorem there exists a uniformly convergent subsequence (y_{nk}) with a limit y^* which is continuous over I_0.

It will be shown that y^* is a solution of (5.11.2) over I_0. If $x > a$ then $x > x_0 + a/n_k$ for all k sufficiently large, so from (5.11.3)

$$y_{nk}(x) = y_0 + \int_{x_0}^{x-a/n_k} f(t, y_{nk}(t))\,dt. \qquad (5.11.4)$$

We cannot let $k \to \infty$ to obtain (5.11.2) immediately; why not? Instead we rewrite (5.11.4) in the form

$$y_{nk}(x) = y_0 + \int_{x_0}^{x} f(t, y_{nk}(t))\,dt - \int_{x-a/n_k}^{x} f(t, y_{nk}(t))\,dt. \quad (5.11.5)$$

Then, by Theorems 1.6.8, 1.6.11,

$$\lim_{k\to\infty} \int_{x_0}^{x} f(t, y_{nk}(t))\,dt = \int_{x_0}^{x} f(t, y^*(t))\,dt.$$

Also

$$\left|\int_{x-a/n_k}^{x} f(t, y_{nk}(t))\,dt\right| \leqslant \int_{x-a/n_k}^{x} M\,dt = Ma/n_k,$$

which tends to zero as $k \to \infty$. Hence, letting $k \to \infty$ in (5.11.5) it follows that

$$y^*(x) = y_0 + \int_{x_0}^{x} f(t, y^*(t))\,dt,$$

so y^* is a solution of (5.11.2), and hence of (5.11.1), over $[x_0, x_0 + a]$.

Similarly there exists a solution of (5.11.1) over $[x_0 - a, x_0]$. This completes the proof.

An example is now given which shows that a differential equation of the form $y' = f(x, y)$, where f is continuous and bounded over a strip S (as defined in Theorem 5.11.1), may not possess a unique solution through (x_0, y_0).

Consider the function f defined by

$$f(x, y) = \min(|y|^{\frac{1}{3}}, 1) \tag{5.11.6}$$

over the strip $S = [-1, 1] \times \mathsf{R}$; let $(x_0, y_0) = (0, 0)$ and $a = 1$. Then f is continuous over S and also is bounded there (it is merely to ensure the latter condition is satisfied that we have defined f by (5.11.6) rather than by $f(x, y) = |y|^{\frac{1}{3}}$). It is easily verified that†

$$y_1(x) = (\operatorname{sgn} x)(\tfrac{2}{3}|x|)^{\frac{3}{2}}$$

and $y_2(x) \equiv 0$ are both solutions of the differential equation over $[-1, 1]$ through $(0, 0)$.

In fact the differential equation possesses an infinite number of solutions through $(0, 0)$; see Hille (1969) pp. 38–39 and Exercise 5.11.1.

Exercises 5.11

1. Let $I = [x_0 - a, x_0 + a]$, and $S = I \times \mathsf{R}$; let $f: S \to \mathsf{R}$ be continuous and bounded over S.

Suppose that the differential equation $y' = f(x, y)$ possesses two distinct solutions y, z through a point (x_0, y_0) in S. Let x_1 be a point in I such that $y(x_1) \neq z(x_1)$ and let c be any real number between $y(x_1)$ and $z(x_1)$.

Prove that there exists a solution k through (x_0, y_0) such that $k(x_1) = c$. Deduce that if there exist two distinct solutions through (x_0, y_0), then there exists an infinite number of such solutions.

2. In the notation of the previous exercise, let Y denote the set of all solutions through (x_0, y_0). Show that Y is

(i) uniformly bounded over I;

(ii) equicontinuous over I;

(iii) a closed subset of $(\mathscr{C}(I), \rho)$ where ρ is the usual supremum metric.

† sgn x is defined to be $+1$ if $x > 0$, -1 if $x < 0$, and 0 if $x = 0$.

6: CONNECTEDNESS

6.1 Introduction and definitions

The intuitive concept underlying connectedness is straightforward; for example $[a, b]$ is a connected subset of the real line, whereas $[a, b] \cup [c, d]$ with $b < c$, is a disconnected subset of the real line. First we give a precise definition for connectedness of metric spaces; as in our study of compactness, we shall initially define the concept for 'whole' metric spaces, and then extend the definition to subsets of metric spaces.

Tentatively we might say that a metric space (X, ρ) is disconnected if there exist two non-empty sets A, B such that

$$X = A \cup B, \quad A \cap B = \varnothing,$$

and that (X, ρ) is connected if no such pair of sets exist. However this implies that all metric spaces with more than one point are disconnected; for, take A to be any proper subset of X and let $B = X - A$. A simple modification of the above tentative definition leads to the following satisfactory form.

DEFINITION 6.1.1. The metric space (X, ρ) is said to be *disconnected* if there exist two non-empty subsets A, B of X such that

$$X = A \cup B, \quad \text{and} \quad A \cap \bar{B} = \varnothing, \quad \bar{A} \cap B = \varnothing. \qquad (6.1.1)$$

If no such sets A, B exist, then (X, ρ) is said to be *connected*.

If (X, ρ) is a disconnected metric space and A, B are a pair of non-empty subsets of X satisfying (6.1.1) then the representation $X = A \cup B$ will be called a *disconnection* of (X, ρ).

A number of equivalent characterizations of disconnectedness (and hence of connectedness) will now be established; these are straightforward.

THEOREM 6.1.1. *Let (X, ρ) be a metric space. Then the following statements are equivalent*:

(i) *(X, ρ) is disconnected*;

(ii) *there exist two non-empty disjoint subsets A, B both open in* (X, ρ) *such that* $X = A \cup B$;

(iii) *there exist two non-empty disjoint subsets A, B both closed in* (X, ρ) *such that* $X = A \cup B$;

(iv) *there exists a proper subset of X which is simultaneously open and closed in* (X, ρ).

Proof. Assume (i). Let $X = A \cup B$ be a disconnection of (X, ρ) so that $A \cap \bar{B} = \varnothing$. Then $A = X - \bar{B}$, so A is open in (X, ρ); likewise B is open in (X, ρ). Clearly $A \cap B = \varnothing$, so (ii) is established.

Trivially (ii) and (iii) are equivalent.

Assume (iii). Since $A = X - B$ and B is closed, A is open; thus A is open and closed in (X, ρ), and (iv) is established.

Assume (iv). Let A be a proper subset of X which is open and closed in (X, ρ), and let $B = X - A$. Then $X = A \cup B$ is a disconnection of X, and (i) is established.

It is clear from the preceding proof that if $X = A \cup B$ is a disconnection of (X, ρ), then A, B are both simultaneously open and closed in (X, ρ).

By negating the statements (i)–(iv) of the above theorem we obtain equivalent characterizations of connectedness.

Connectedness is preserved by homeomorphic mappings. More precisely we have:

THEOREM 6.1.2. *Let (X, ρ), (X', ρ') be homeomorphic metric spaces; (X, ρ) is connected if and only if (X', ρ') is connected.*

Proof. Let $f:(X, \rho) \to (X', \rho')$ be a homeomorphism. Suppose (X', ρ') is disconnected and let A' be a proper subset of X' which is simultaneously open and closed in (X', ρ'). Then $f^{-1}(A')$ is both open and closed in (X, ρ) and is also a proper subset of (X, ρ); (X, ρ) is therefore disconnected. Hence (X, ρ) is disconnected if and only if (X', ρ') is disconnected, and so the theorem follows.

EXERCISES 6.1

1. Show that any discrete metric space with more than one point is disconnected.

2. Show that (Q, d) is disconnected.

3. Let (X, ρ) be a connected space and $f:(X, \rho) \to (\mathsf{R}, \mathsf{d})$ be continuous and such that, for any y in X for which $f(y) = 0$ there exists a neighbourhood $S(y, \varepsilon)$ such that $f(x) = 0$ for all x in $S(y, \varepsilon)$; prove that either $f(x) = 0$ for all x in X, or $f(x) \neq 0$ for all x in X.

4. If (X, ρ) is connected and X contains more than one point, prove that X must be infinite. [In fact X must be uncountably infinite: see Exercise 6.4.3.]

5. Let X be a non-empty set and let $x, y \in X$. The finite sequence $(Y_1, \ldots, Y_n)$ of subsets of X is said to form a *finite* (or *simple*) *chain* joining x to y if $x \in Y_1$, $y \in Y_n$ and

$$Y_1 \cap Y_2 \neq \varnothing, \quad Y_2 \cap Y_3 \neq \varnothing, \ldots, Y_{n-1} \cap Y_n \neq \varnothing.$$

Show, by means of the following steps, that a metric space is connected if and only if, for any open covering $\mathscr{Y} = \{Y_\lambda : \lambda \in \Lambda\}$ of (X, ρ), every pair of points in X can be joined by a finite chain of members of $\mathscr{Y}$.

Suppose (X, ρ) is connected. Let $x \in X$ and let $H(x)$ be the set of all points in (X, ρ) which can be joined to x by a finite chain of members of $\mathscr{Y}$.

(i) Let $h \in H(x)$ and $(Y_1, \ldots, Y_n)$ be a finite chain from h to x. Show that $Y_1 \subseteq H(x)$; deduce that $H(x)$ is open.

(ii) Suppose, if possible, that $X - H(x) \neq \varnothing$ and let $g \in X - H(x)$; let $g \in Y_{\lambda_0}$. Show that $Y_{\lambda_0} \cap H(x) = \varnothing$ and deduce that $X - H(x)$ is open. Hence show that $H(x) = X$.

Obtain the converse result by a contradiction argument.

6. Let X be a subset of R and d_1, d_2 be the Euclidean metrics for R, R^2 respectively. Given a function $f:(X, \mathsf{d}_1) \to (\mathsf{R}, \mathsf{d}_1)$, define the function $F:(X, \mathsf{d}_1) \to (F(X), \mathsf{d}_2)$, where $F(X) \subseteq \mathsf{R}^2$, by $F(x) = (x, f(x))$ for all x in X. Prove that F is a homeomorphism if and only if f is continuous.

If f is continuous show that $F(X)$ is compact, or connected, if and only if X is compact, or connected (respectively).

6.2 Connected and disconnected subsets

DEFINITION 6.2.1. A non-empty subset Y of a metric space (X, ρ) is said to be a *connected* subset of (X, ρ) if (Y, ρ_Y) is connected and to be a *disconnected* subset of (X, ρ) if (Y, ρ_Y) is disconnected.

Let (X, ρ) be any metric space and $Y \subseteq X$; if Y contains exactly one point, then it is a connected subset.

Let (X, ρ) be a discrete metric space and $Y \subseteq X$; if Y contains more than one point, then it is a disconnected subset.

By Definition 6.2.1, Y is a disconnected subset of (X, ρ) if and only if there exist non-empty subsets A, B of Y such that

$$Y = A \cup B; \quad \bar{A}^Y \cap B = \varnothing, \quad A \cap \bar{B}^Y = \varnothing, \qquad (6.2.1)$$

where $\bar{A}^Y$, $\bar{B}^Y$ denote the closures of A, B in (Y, ρ_Y). (It should be remembered that, if (X, ρ) is any metric space and $Z \subseteq Y \subseteq X$, then $\bar{Z}^Y = \bar{Z}^X \cap Y$ (see Lemma 2.8.2) so that $\bar{Z}^Y$, $\bar{Z}^X$ are not necessarily equal.) However, since our basic metric space is (X, ρ) it would be more convenient if we did not have to introduce the closures of A, B with respect to (Y, ρ_Y). It happens that this can easily be arranged; we have the following result.

THEOREM 6.2.1. *Y is a disconnected subset of (X, ρ) if and only if there exist non-empty sets A, B such that*

$$Y = A \cup B; \quad \bar{A}^X \cap B = \varnothing, \quad A \cap \bar{B}^X = \varnothing. \qquad (6.2.2)$$

Proof. First assume that there exist sets A, B which satisfy (6.2.2). Since $\bar{A}^X \supseteq \bar{A}^Y$, $\bar{B}^X \supseteq \bar{B}^Y$, (6.2.1) follows and thus Y is a disconnected subset of (X, ρ).

Conversely, assume that Y is disconnected so there exist non-empty sets A, B which satisfy (6.2.1). Since $\bar{A}^Y = \bar{A}^X \cap Y$ it follows that

$$\bar{A}^Y \cap B = (\bar{A}^X \cap Y) \cap B = \bar{A}^X \cap B,$$

and similarly

$$A \cap \bar{B}^Y = A \cap \bar{B}^X,$$

so (6.2.1) implies (6.2.2), and the theorem is proved.

DEFINITION 6.2.2. If A, B are two subsets of (X, ρ) such that

$$\bar{A} \cap B = \varnothing, \quad A \cap \bar{B} = \varnothing,$$

then A, B are said to be *separated*.

Thus a subset Y of (X, ρ) is disconnected if and only if it can be expressed as the union of two non-empty separated sets.

Let Y be a disconnected subset of (X, ρ) and A, B be two non-empty sets such that

$$Y = A \cup B, \quad \bar{A}^Y \cap B = \varnothing, \quad A \cap \bar{B}^Y = \varnothing,$$

so

$$\bar{A}^X \cap B = \varnothing, \quad A \cap \bar{B}^X = \varnothing.$$

Of course, although A, B are both open and closed relative to (Y, ρ_Y), so $A = \bar{A}^Y$, $B = \bar{B}^Y$, it does not follow that A, B are both open and closed relative to (X, ρ). For example, let $X = \mathsf{R}$, $Y = [0, 1) \cup [2, 3)$ and ρ be the Euclidean metric; then we may take $A = [0, 1)$, $B = [2, 3)$. It follows that

$$\bar{A}^X = [0, 1] \neq A, \quad \bar{B}^X = [2, 3] \neq B,$$

so that neither A nor B is both open and closed relative to (X, ρ).

THEOREM 6.2.2. *If A, B are non-empty separated sets of (X, ρ) and Y is a connected subset of (X, ρ) such that $Y \subseteq A \cup B$, then either $Y \subseteq A$ or $Y \subseteq B$.*

Proof. Suppose, if possible, $Y \cap A \neq \varnothing$ and $Y \cap B \neq \varnothing$; it will then be shown that Y can be expressed as the union of two non-empty separated sets namely $Y \cap A$, $Y \cap B$. By Theorem 2.7.3,

$$\overline{Y \cap A} \subseteq \bar{Y} \cap \bar{A}$$

so

$$\overline{(Y \cap A)} \cap (Y \cap B) \subseteq (\bar{Y} \cap \bar{A}) \cap (Y \cap B) = Y \cap (\bar{A} \cap B) = \varnothing,$$

since $\bar{A} \cap B = \varnothing$. Similarly

$$(Y \cap A) \cap \overline{(Y \cap B)} = \varnothing,$$

and so, by Theorem 6.2.1, Y is disconnected. This gives the required contradiction.

Let A, B be connected subsets of (X, ρ) which are not separated. Suppose, if possible, that $A \cup B = C \cup D$ is a disconnection of $A \cup B$. Then by Theorem 6.2.2, $A \subseteq C$ or $A \subseteq D$; for definiteness assume the former. Then $B \subseteq D$ (for $B \subseteq C$ implies that $D = \varnothing$). Hence A, B are subsets of separated sets and so must also be separated, which is impossible. Thus the union of two connected sets which are not separated is also connected; this can be generalized to the union of any collection of connected subsets of (X, ρ).

THEOREM 6.2.3. *If $\{Y_\lambda : \lambda \in \Lambda\}$ is a collection of connected subsets of (X, ρ) containing at least one member Y_{λ_0} which is not separated from any of the remaining Y_λ, then*

$$Y = \bigcup_{\lambda \in \Lambda} Y_\lambda$$

is also a connected set.

Proof. Suppose, if possible, that Y is disconnected; let $Y = A \cup B$ be a disconnection. Then $Y_{\lambda_0} \subseteq A \cup B$ so, by Theorem 6.2.2, $Y_{\lambda_0} \subseteq A$ or $Y_{\lambda_0} \subseteq B$; suppose the former. Also, for any λ ($\neq \lambda_0$) in Λ, as we have seen $Y_{\lambda_0} \cup Y_\lambda$ is connected and

$$Y_{\lambda_0} \cup Y_\lambda \subseteq A \cup B;$$

therefore $Y_{\lambda_0} \cup Y_\lambda \subseteq A$ (since $Y_{\lambda_0} \cup Y_\lambda \subseteq B$ is impossible). Thus

$$\bigcup_{\lambda \in \Lambda} \{Y_{\lambda_0} \cup Y_\lambda\} \subseteq A,$$

that is, $Y \subseteq A$, and $B = \varnothing$. This gives the required contradiction.
The following are important special cases of the last result.

COROLLARY 6.2.1. *If $\{Y_\lambda : \lambda \in \Lambda\}$ is a family of connected subsets of (X, ρ) such that*

$$\bigcap_{\lambda \in \Lambda} Y_\lambda = \varnothing,$$

then

$$Y = \bigcup_{\lambda \in \Lambda} Y_\lambda$$

is also a connected subset.

COROLLARY 6.2.2. *If (A_n) is a sequence of connected subsets of (X, ρ) such that $A_n \cap A_{n+1} \neq \varnothing$ for $n = 1, 2, \ldots$, then*

$$\bigcup_{i=1}^{n} A_i, \quad \bigcup_{i=1}^{\infty} A_i$$

are also connected subsets.

We can deduce the following result from Theorem 6.2.3.

THEOREM 6.2.4. *Let Y be a connected subset of (X, ρ). If Z is any subset of X such that $Y \subseteq Z \subseteq \bar{Y}$, then Z is also a connected subset of (X, ρ).*

Proof. We may write Z as

$$Z = Y \cup \left\{ \bigcup_{y \in Z - Y} \{y\} \right\};$$

for each y in $Z-Y$, Y is not separated from $\{y\}$ (since each such y is a limit point of Y) and so, by Theorem 6.2.3, Z is connected.

EXERCISES 6.2

1. In a metric space (X, ρ) prove that

(i) if A, B are separated, and $A' \subseteq A$, $B' \subseteq B$ then A', B' are separated;

(ii) if A, B are separated and A, C are separated then $A, B \cup C$ are separated;

(iii) if A, B are both closed (or both open) then $A-B$, $B-A$ are separated.

2. Let A, B be non-empty subsets of (X, ρ). If $\rho(A, B) > 0$ prove that A, B are separated. Show that the converse is false.

3. Establish Theorem 6.2.4 by the following contradiction argument. Suppose $Z = A \cup B$ is a disconnection of Z; then either $Y \subseteq A$ or $Y \subseteq B$. Show that the former implies $B = \varnothing$ and the latter implies $A = \varnothing$.

4. Show that any space (X, ρ) is connected if and only if, for every pair of points of X, there exists a connected subspace of (X, p) which contains both.

6.3 The connected sets of the real line

A subset I of R is called an interval if and only if it is a set of one of the following forms

$$(a, b),\quad [a, b),\quad (a, b],\quad [a, b],$$
$$(-\infty, b),\quad (-\infty, b],\quad (a, \infty),\quad [a, \infty),\quad (-\infty, \infty),$$

where $a, b \in$ R (and $a < b$); it is easily seen, using the least upper bound axiom, that this is equivalent to saying that I is an interval if and only if whenever $x, z \in I$ and $x < y < z$, then $y \in I$.

As we should expect, the connected subsets of the real line are precisely the intervals.

THEOREM 6.3.1. *A non-empty subset of* (R, d) *is connected if and only if it is an interval.*

Proof. Assume, first, that Y is a connected subset of (R, d) and suppose, if possible, that Y is not an interval. Then there exist real numbers x, y, z such that $x < y < z$, $x, z \in Y$, but $y \notin Y$. Let

$$A = Y \cap (-\infty, y), \quad B = Y \cap (y, \infty).$$

Then, using Theorem 2.8.2, it follows that A, B are both open relative to (Y, d_Y). Furthermore, A, B are non-empty and $Y = A \cup B$. Hence (Y, d_Y) is disconnected, which gives a contradiction. Thus Y is an interval.

Next assume, conversely, that Y is an interval (which may be the whole of R). Suppose, if possible, that (Y, d_Y) is disconnected, and let $Y = A \cup B$ be a disconnection; we shall use the bisection method to obtain a contradiction. Take arbitrary points a_1 in A, b_1 in B and suppose that $a_1 < b_1$. Bisect the interval (a_1, b_1); then the midpoint belongs to Y, and so belongs to one and only one of the sets A, B. Therefore, exactly one of the half-intervals will have its left end-point in A and its right end-point in B. Call this interval (a_2, b_2); bisect it and proceed as before. Thus we may define, by induction, two monotonic sequences (a_n), (b_n) which, by the usual argument†, converge to a common limit ξ, say. Clearly $\xi \in Y$; since A, B are closed relative to (Y, d_Y), ξ belongs to them both. This gives the required contradiction.

Theorem 6.3.1 implies, of course, that the real line, with Euclidean metric, is connected.

EXERCISES 6.3

1. By the following argument show that $(\mathsf{R}^m, \mathsf{d})$, where $m \geqslant 2$, is connected.

Suppose, if possible, there exists a proper subset Y of R^m which is open and closed. Let $x \in Y$ and $y \in \mathsf{R}^m - Y$; let L be the (infinite) straight line through x, y. Show that $L \cap Y$ is both open and closed in (L, d_L). Hence obtain a contradiction.

With the aid of the example in §3.4 deduce that the metric spaces defined in §2.2(iii), (iv) are also connected.

2. Let $(\mathscr{C}(I), \rho)$ be the metric space defined in §2.2(viii).

Let $x \in \mathscr{C}(I)$ and define $[x] = \{\lambda x : -1 \leqslant \lambda \leqslant 1\}$, so $[x]$ is a

† See, for example, the proof of Proposition 5.3.1.

subset of $\mathscr{C}(I)$. Show that $[x]$, with supremum metric, is homeomorphic to $[-1, 1]$ with Euclidean metric. Writing

$$\mathscr{C}(I) = \bigcup_{x \in \mathscr{C}(I)} [x],$$

deduce that $(\mathscr{C}(I), \rho)$ is connected.

Modify this argument to show that the metric spaces defined in §2.2(vi), (vii), (ix), (x) are all connected.

6.4 The generalization of the intermediate value theorem

The following is an important result of real variable theory.

PROPOSITION 6.4.1 (The intermediate value theorem). *If f: $[a, b] \to \mathrm{R}$ is continuous over $[a, b]$, then for any y such that $f(a) \leqslant y \leqslant f(b)$ there exists c in $[a, b]$ such that $f(c) = y$.*

More generally if I is any interval of the real line, if $f: I \to \mathrm{R}$ is continuous, if $h, k \in f(I)$ and $h < y < k$, then there exists c in I such that $f(c) = y$; thus f maps an interval into an interval.

This is generalized to arbitrary metric spaces as follows.

THEOREM 6.4.1. *Let (X, ρ), (X', ρ') be two metric spaces and Y be a connected subset of (X, ρ); if f is a continuous mapping of (Y, ρ_Y) into (X', ρ') then $f(Y)$ is a connected subset of (X', ρ').*

Proof. Let $Y' = f(Y)$. Suppose, if possible, that Y' is a disconnected subset of (X', ρ'). Then there exists a proper subset, A' say, of Y' which is simultaneously open and closed in $(Y', \rho'_{Y'})$. Since $f:(Y, \rho_Y) \to (Y', \rho'_{Y'})$ is continuous it follows from Theorem 3.2.1 that $f^{-1}(A')$ is simultaneously open and closed in (Y, ρ_Y). Moreover since f maps Y onto $f(Y)$ it follows that $f^{-1}(A')$ is a proper subset of Y; hence (Y, ρ_Y) is disconnected, which gives the required contradiction.

This result is summarized by saying that 'a continuous function maps a connected set into a connected set'. It implies, of course, Theorem 6.1.2. Note however that a continuous function does not necessarily map a disconnected set into a disconnected set.

As an application of the above result we derive another characterization of connectedness. For the remainder of this section let $X_0 = \{0, 1\}$ and let ρ_0 denote the standard discrete metric on X_0: we shall call (X_0, ρ_0) the *discrete two-point* metric space.

THEOREM 6.4.2. *Let (X, ρ) be a metric space. Then the following statements are equivalent:*

(i) *(X, ρ) is disconnected;*

(ii) *there exists a continuous mapping f of (X, ρ) onto (X_0, ρ_0).*

Proof. Assume (i); let $X = A \cup B$ be a disconnection of (X, ρ). Define a mapping $f: X \to X_0$ by

$$f(x) = \begin{cases} 0 & \text{if } x \in A, \\ 1 & \text{if } x \in B; \end{cases}$$

clearly this is a mapping of X onto X_0.

It remains to prove that $f:(X, \rho) \to (X_0, \rho_0)$ is continuous. Since (X_0, ρ_0) is a discrete metric space, every subset in it is open; there are precisely four subsets, namely $\varnothing$, X, $\{0\}$, $\{1\}$. But $f^{-1}(\varnothing) = \varnothing$, $f^{-1}(X_0) = X$, and $\varnothing$, X are open in (X, ρ); furthermore $f^{-1}(\{0\}) = A$, $f^{-1}(\{1\}) = B$ and A, B are both open in (X, ρ) since $A \cup B$ is a disconnection of (X, ρ). Hence f is continuous, and thus (ii) is proved.

Assume (ii) and suppose, if possible, that (X, ρ) is connected. Then by Theorem 6.4.1 (X_0, ρ_0) is connected, which is impossible, so (i) is proved.

This result may be reformulated as follows.

COROLLARY 6.4.1. *Let (X, ρ) be a metric space. Then the following statements are equivalent:*

(i) *(X, ρ) is connected;*

(ii) *the only continuous mappings of (X, ρ) into (X_0, ρ_0) are the constant mappings, (that is, the mappings f, g defined by $f(x) = 1$ for all x in X, $g(x) = 0$ for all x in X).*

EXERCISES 6.4

1. Show, by means of the following outline, that the metric spaces (X, ρ), (X, ρ') defined in Exercise 2.2.2 are connected if and only if the spaces (X_i, ρ_i), $i = 1, \ldots, m$ are all connected.

For the 'only if' part use Exercise 3.2.8.

For the 'if' part, first assume that $m = 2$. Let $x_1 \in X_1$, $x_2 \in X_2$; show that $X_1 \times \{x_2\}$, $\{x_1\} \times X_2$ are connected subsets of (X, ρ) so that

$$(X_1 \times \{x_2\}) \cup (\{x_1\} \times X_2)$$

is connected. Take the union over all x_1 in X_1 to show that (X, ρ) is connected. Use induction to obtain the conclusion for any m.

2. In the notation of Exercise 6.1.6 show that, if $F(X)$ is a connected subset of $(\mathsf{R}^2, \mathsf{d})$, then $f(X)$ is an interval. [In order that $F(X)$ be connected it is not necessary for f to be continuous—see the example given immediately after the proof of Theorem 6.5.1.]

3. Let (X, ρ) be a connected metric space containing more than one point. Let $a \in X$; by considering the function $f:(X, \rho) \to (\mathsf{R}, \mathsf{d})$ defined by $f(x) = \rho(a, x)$ for all x in X, prove that X must be uncountably infinite.

4. Let $I = [0, 1]$ and let the function $f:(I, \mathsf{d}) \to (I, \mathsf{d})$ be continuous. By means of the following contradiction argument show that there exists x in I such that $f(x) = x$.

Let $A = \{x:x < f(x)\}$, $B = \{x:x > f(x)\}$; show that A, B are open subsets of (I, d). Assume that $f(0) > 0, f(1) < 1$ (for otherwise the result is immediate); if there is no x such that $f(x) = x$ show that $A \cup B$ is a disconnection of (I, d).

5. By using Theorem 6.4.2 and a contradiction argument prove that a metric space (X, ρ) is connected if every continuous function $f:(X, \rho) \to (\mathsf{R}, \mathsf{d})$ has the intermediate value property; that is, if $y_1, y_2 \in f(X)$ and y is any real number between y_1 and y_2, then there exists x in X such that $f(x) = y$. [Observe that the result of this exercise is a converse to the result of Theorem 6.4.1.]

6.5 Pathwise connectedness

In §6.1 we defined the concept of connectedness; this definition was motivated by the idea that a space is connected if it is 'all of one piece'. Another starting point would be to regard a space (X,p) as connected if every pair of points in X could be joined by a continuous path lying entirely in X. In the present section we formalize this idea and call such spaces pathwise connected; we then show how this concept is related to that concept of connectedness already defined.

DEFINITION 6.5.1. Let (X, ρ) be a metric space and let the function $f:([0, 1], \mathsf{d}) \to (X, \rho)$ be continuous; the range of f, that is, the set $f([0, 1])$, is called a *path from* $f(0)$ *to* $f(1)$ *in* (X, ρ).

In this definition $[0, 1]$ may, of course, be replaced by any closed bounded interval $[a, b]$.

DEFINITION 6.5.2. A metric space (X, ρ) is said to be *pathwise connected* if, for every pair of points x, y in X, there exists a path from x to y lying entirely in (X, ρ). A non-empty subset Y of (X, ρ) is said to be pathwise connected if (Y, ρ_Y) is pathwise connected.

The term arcwise connected is used by some authors instead of pathwise connected, but by others it is used in a slightly different sense.

If a, b, c are three points in a metric space (X, ρ) and f, g are mappings of $([0, 1], \mathsf{d})$ into (X, ρ) whose ranges are paths from a to b and from b to c respectively, then clearly there exists a mapping h of $([0, 1], \mathsf{d})$ into (X, ρ) whose range is a path from a to c, and which path contains b. This fact will be used in the proof of Theorem 6.5.2.

THEOREM 6.5.1. *A pathwise connected metric space is connected.*

Proof. Let (X, ρ) be a pathwise connected metric space and let $x \in X$; then for each y in X there exists a continuous function $f_y:([0, 1], \mathsf{d}) \to (X, \rho)$ such that $f_y(0) = x$, $f_y(1) = y$. Since $[0, 1]$ is connected, so also is $f_y([0, 1])$; moreover

$$x \in \bigcap_{y \in X} f_y([0, 1]) \tag{6.5.1}$$

so the right side of (6.5.1) is not empty. Hence by Corollary 6.2.1 it follows that

$$\bigcup_{y \in X} f_y([0, 1])$$

is connected in (X, ρ); that is (X, ρ) is connected.

However a connected space is not necessarily pathwise connected; this is shown by the following example.

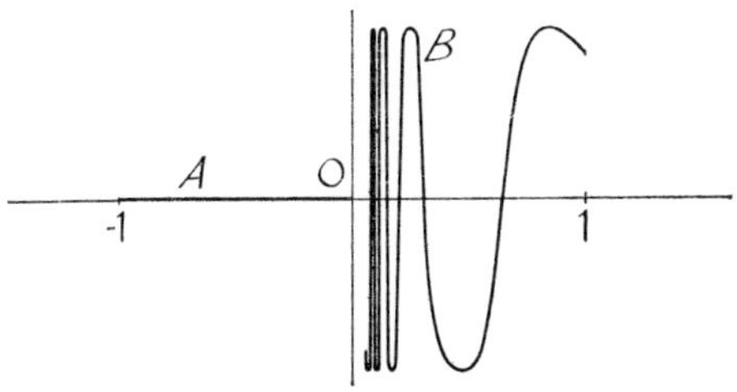

In $(\mathsf{R}^2, \mathsf{d})$ let

$$A = \{(x, 0): -1 \leqslant x \leqslant 0\}, \quad B = \{(x, \sin x^{-1}): 0 < x \leqslant 1\}.$$

The sets A, B are both connected (the latter since it is the image under a continuous function of $(0, 1]$). Moreover every point $(0, y)$ where $-1 \leqslant y \leqslant 1$ is a limit point of B so that $A \cap \bar{B} = \{(0, 0)\}$; hence A, B are not separated, so $A \cup B$ is connected.

However $A \cup B$ is not pathwise connected; this fact, although intuitively reasonable, is somewhat tricky to prove. We use a contradiction argument.

Thus suppose, if possible, that there exists a continuous function f which maps $([0, 1], \mathsf{d})$ into $(A \cup B, \rho)$ and $f(0) = (0, 0)$, $f(1) = (1, \sin 1)$. Let $f(t) = (x(t), y(t))$ for all t in $[0, 1]$, so that

$$y(t) = \begin{cases} 0 & \text{if } -1 \leqslant x(t) \leqslant 0, \\ \sin\{x(t)\}^{-1} & \text{if } 0 < x(t) \leqslant 1. \end{cases}$$

Let t_0 be the largest root in $[0, 1]$ of $x(t) = 0$; why does such a t_0 exist? Then $t_0 < 1$ since $x(1) = 1$. By the continuity of x it follows that $x(t) > 0$ over $(t_0, 1]$. Using the continuity of y at t_0, and the fact that $y(t_0) = 0$ it follows that there exists $\delta > 0$ (and such that $t_0 + \delta < 1$) for which $|y(t)| < 1$ for all t in $(t_0, t_0 + \delta)$. Since $x(t_0 + \delta) > 0$ there exists a positive integer n such that

$$x(t_0+\delta) > \frac{1}{(2n+\frac{1}{2})\pi} > x(t_0)$$

so, by the intermediate value theorem, there exists t_1 in $(t_0, t_0+\delta)$ for which

$$x(t_1) = \frac{1}{(2n+\frac{1}{2})\pi};$$

but then $y(t_1) = 1$ which gives a contradiction.

Although a connected set is not necessarily pathwise connected, we do have the following result for open sets of Euclidean space.

THEOREM 6.5.2. *In* $(\mathsf{R}^m, \mathsf{d})$ *any open set* X *is connected if and only if it is pathwise connected.*

Proof. By the previous theorem, if X is pathwise connected then it is connected.

Now suppose that X is an open connected set in $(\mathsf{R}^m, \mathsf{d})$. Fix x in X and let A be the set of all a in X such that there exists a path from x to a lying in X; clearly $x \in A$ so $A \neq \emptyset$. Let $B = X - A$; then $A \cap B = \emptyset$ and $X = A \cup B$. It will be shown that A, B are both open; from this it follows that $B = \emptyset$, for otherwise $X = A \cup B$ would be a disconnection of X.

Let $a \in A$; join x to a by a path P lying in X. Since X is open there exists $\delta > 0$ such that $S(a, \delta) \subseteq X$; but since every y in $S(a, \delta)$ can be joined to a by a straight line in X, y can be joined to x by a path in X. Thus $y \in A$ so $S(a, \delta) \subset A$, and therefore A is open. Next let $b \in B$; then there exists $\delta > 0$ such that $S(b, \delta) \subseteq X$. No point of $S(b, \delta)$ can be joined to x by a path lying entirely in X, for otherwise it would be possible to join b and x by a path in X. Hence every point of $S(b, \delta)$ is in B, that is $S(b, \delta) \subseteq B$, so B is open. This concludes the proof.

The essential feature of $(\mathsf{R}^m, \mathsf{d})$ that is necessary for the proof of Theorem 6.5.2 is that any open sphere of R^m is pathwise connected. Abstraction of this idea leads us to a generalization of the result of Theorem 6.5.2 to any connected metric space (X, ρ) having the property that for every point x in X there exists an arbitrarily small neighbourhood of x which is pathwise connected (any space with this property is said to be *locally pathwise connected*). For further discussion of this the reader is referred to Gleason (1966), p. 284.

Finally we record the following result, the proof of which is left to the reader.

THEOREM 6.5.3. *If (X, ρ), (X', ρ') are homeomorphic then (X, ρ) is pathwise connected if and only if (X', ρ') is pathwise connected.*

EXERCISES 6.5

1. Show that the image of a pathwise connected set under a continuous function is itself pathwise connected.

2. If $\{Y_\lambda : \lambda \in \Lambda\}$ is a collection of pathwise connected subsets of (X, ρ) such that

$$\bigcap_{\lambda \in \Lambda} Y_\lambda \neq \emptyset,$$

then prove that

$$Y = \bigcup_{\lambda \in \Lambda} Y_\lambda$$

is also a pathwise connected subset. [This result is an analogue for pathwise connectedness of Corollary 6.2.1.]

3. By means of the following steps give an alternative proof of Theorem 6.5.1.

Let (X, ρ) be a pathwise connected metric space and suppose, if possible, that there is a disconnection $X = A \cup B$ of (X, ρ). Let $a \in A$, $b \in B$ and let $f:([0, 1], \mathrm{d}) \to (X, \rho)$ be continuous and such that the range of f is a path from a to b. Show that $f^{-1}(A), f^{-1}(B)$ are both non-empty subsets of $[0, 1]$. Hence obtain a contradiction.

4. By means of the following steps give an alternative proof of Theorem 6.5.2.

Express the open set X as the union of a collection of open spheres. Let $a, b \in X$; using Exercise 6.1.5 show that there exists a path from a to b.

6.6 The components of a disconnected space

Although a metric space may not be connected, it can be split up into disjoint sets each of which is connected. How this is done is now described.

Let (X, ρ) be a metric space and $x \in X$. Denote by $C(x)$ the union of all connected subsets of (X, ρ) which contain x; thus $C(x)$ contains all connected subsets which contain x. Moreover, by Corollary 6.2.1, $C(x)$ is a connected subset of (X, ρ).

DEFINITION 6.6.1. The set $C(x)$ so defined is called the *connected component* (or *component*) of x in (X, ρ).

If $y \in C(x)$, then $C(y) \cap C(x) \neq \varnothing$ so, by §6.2, $C(y) \cup C(x)$ is a connected subset containing x; hence $C(y) \cup C(x) = C(x)$, so $C(y) \subseteq C(x)$. Also since $y \in C(x)$, it follows that $C(x) \subseteq C(y)$; therefore $C(x) = C(y)$. In a similar fashion if $y \notin C(x)$, it follows that $C(x) \cap C(y) = \varnothing$.

We define an equivalence relation in X, to be denoted by $\sim$, such that $x \sim y$ if and only if $C(x) = C(y)$. The components of (X, ρ) are precisely the equivalence classes of this equivalence relation.

THEOREM 6.6.1. *Let (X, ρ) be a metric space. Then*

(i) *each point x of X belongs to exactly one component;*

(ii) *each connected subset of (X, ρ) is contained in exactly one component of (X, ρ);*

(iii) *each connected subset (X, ρ) which is open and closed in (X, ρ) is a component of (X, ρ);*

(iv) *each component of* (X, ρ) *is closed.*

Proof. (i) and (ii) follow immediately from the comments preceding the statements of the theorem.

For (iii), let A be a connected subset which is open and closed in (X, ρ). Let $x \in A$, so $A \subseteq C(x)$; then (by Theorem 2.8.1) A is also open and closed in $(C(x), \rho_{C(x)})$. Hence $A = C(x)$.

For (iv), use Theorem 6.2.4.

Note that the components of (X, ρ) are not necessarily open. For let

$$X = \{0\} \cup \left\{ \bigcup_{n=1} \{1/n\} \right\},$$

and let ρ be the Euclidean metric; then $\{0\}$ is a component of (X, ρ) but is not open in (X, ρ).

If, however, it is assumed that (X, ρ) is *locally connected*, that is, for every x in X there exists an arbitrarily small neighbourhood of x which is connected, then every component of (X, ρ) is open.

If a metric space has the property that its components are all its distinct points, then it is said to be *totally disconnected.*

For further discussion of these last two comments see, for example, Dieudonné (1960), §3.19, Simmons (1963), §§33, 34, or Gleason (1966), pp. 282–3.

EXERCISES 6.6

1. Show that any pair of distinct components are separated.

2. Let (X, ρ) be a compact metric space; if the components of (X, ρ) are all open, show that there are at most a finite number of components.

3. If (X, ρ) has only a finite number of components, then prove that each component is both open and closed.

4. Let Y be a component of (X', ρ') and $f:(X, \rho) \to (X', \rho')$ be continuous; prove that $f^{-1}(Y)$ is the union of components of (X, ρ). Give an example to show that $f^{-1}(Y)$ is not necessarily a component of (X, ρ).

5. Let Y be an open subset of $(\mathsf{R}^m, \mathsf{d})$; show that every component of (Y, d) is an open subset of $(\mathsf{R}^m, \mathsf{d})$.

7: FURTHER TOPICS

A. EXTENSION THEOREMS

7.1 Introduction; limits

Let X, X' be non-empty sets and $Y \subseteq X$; let $f: Y \to X'$ be a given function and let $F: X \to X'$ be a function such that $f(y) = F(y)$ for all y in Y. Then F is called an *extension* of f. It is trivial that an extension F of f always exists; for example let F be defined by

$$F(x) = \begin{cases} f(x) & \text{if } x \in Y \\ x_0' & \text{if } x \notin Y \end{cases}$$

where x_0' is any element of X'. The problem of finding an extension F becomes non-trivial if we require that F should have certain properties. For example if X, X' have appropriate structure we might require that F be bounded, or continuous, or differentiable, etc. In particular we often require that F should have some property also possessed by f. Thus if f is continuous we might ask whether there exists a continuous extension F of f. Another extension problem is that of analytic continuation; in this context analytic extensions are sought of functions $f: Y(\subset \mathbb{C}) \to \mathbb{C}$ which are analytic.

Throughout this chapter, unless the contrary is stated, it should be assumed that the Euclidean metric is associated with $\mathbb{R}$.

First we start with some preliminary results. These were contained in Exercise 3.1.4; for completeness they are repeated here.

LEMMA 7.1.1. *Let (X, ρ), (X', ρ') be two metric spaces and f, g be two continuous mappings of (X, ρ) into (X', ρ').*

(i) *If $f(x) = g(x)$ for all x in A, then $f(x) = g(x)$ for all x in $\bar{A}$; thus if $B = \{x: f(x) = g(x)\}$, then B is closed.*

(ii) *If $X' = \mathbb{R}$, ρ' is the Euclidean metric and $C = \{x: f(x) \geqslant g(x)\}$, then C is closed.*

Thus if it is known that $f(x) = g(x)$ on a set which is everywhere dense in (X, ρ), then it follows that $f(x) = g(x)$ everywhere so $f = g$.

This fact is of considerable use, for example, in solving functional equations such as

$$f(x+y) = f(x)+f(y), \quad x, y \in \mathsf{R} \tag{7.1.1}$$

$$f(x+y) = f(x).f(y), \quad x, y \in \mathsf{R} \tag{7.1.2}$$

$$f(x.y) = f(x)+f(y), \quad x, y \in \mathsf{R}^+ \tag{7.1.3}$$

where $f\colon \mathsf{R} \to \mathsf{R}$ (or $f\colon \mathsf{R}^+ \to \mathsf{R}$ in (7.1.3)) is continuous, and R^+ is the set of all positive reals. It can be shown (see Exercise 7.1.1) that the solution of (7.1.1) is of the form $f(x) = cx$ for all rational x, so is of this form for all real x. Likewise it can be shown that the solution of (7.1.2) is either of the form $f(x) = 0$ or of the form $f(x) = a^x$ (where $a > 0$) for all rational x, so is of this form for all real x.

For (7.1.3) it can be shown that $f(x) = 0$ or $f(e^x) = xf(e)$ for all rational x, and hence for all real x; thus $f(x) = 0$ or $f(x) = \log_a x$ (where $a > 0$) for all real $x > 0$.

When studying functions on the real line, it is customary to introduce the idea of a limit of a function as the independent variable tends to some fixed real number; this idea is easily generalized to any metric space.

DEFINITION 7.1.1. Let (X, ρ), (X', ρ') be two metric spaces and $Y \subseteq X$; let f be a function of (Y, ρ_Y) into (X', ρ'). Let x be a limit point of Y and $x' \in X'$. If, given any $\varepsilon > 0$, there exists $\delta > 0$ such that

$$\rho'(f(y), x') < \varepsilon \text{ for all } y \text{ in } Y \text{ such that } 0 < \rho(y, x) < \delta, \tag{7.1.4}$$

we say that $f(y)$ *tends to* x' *as* y *tends to* x *in* Y, and write $f(y) \to x'$ as $y \to x$ in Y or

$$\lim_{y \to x,\, y \in Y} f(y) = x'.$$

We make some remarks concerning this definition.

(i) We may omit reference to the set Y if no confusion can arise.

(ii) The limit described in the definition is unique.

(iii) The requirement (7.1.4) is equivalent to

$$f(Y \cap (S(x, \delta) - \{x\})) \subseteq S'(x', \varepsilon),$$

or

$$Y \cap (S(x, \delta) - \{x\}) \subseteq f^{-1}(S'(x', \varepsilon)),$$

where S' denotes a sphere in (X', ρ').

(iv) The existence and value of a limit of $f(y)$ as $y \to x$ in Y is independent of the value of f at x, even if f is defined there—which is not necessary for the limit to exist.

(v) The purpose of requiring that x be a limit point of Y is to ensure that, for any $\delta > 0$,

$$Y \cap (S(x, \delta) - \{x\}) \neq \varnothing.$$

If x is not a limit point of Y, there exists $\delta_0 > 0$ such that

$$Y \cap (S(x, \delta_0) - \{x\}) = \varnothing$$

and hence, given any $\varepsilon > 0$ and *any* x' in X',

$$f(Y \cap (S(x, \delta_0) - \{x\})) \subseteq S'(x', \varepsilon),$$

so $f(y) \to x'$; such a situation is therefore excluded by our definition.

(vi) If the function f is continuous over Y then, for any limit point x of Y, $f(y)$ tends to the limit $f(x)$ as y tends to x.

We state a result, the proof of which is left as an exercise.

LEMMA 7.1.2. *In the notation of Definition* 7.1.1, *if* $f(y) \to x'$ *as* $y \to x$ *in* Y, *then* $x' \in \overline{f(Y)}$.

EXERCISES 7.1

1. The mapping $f\colon \mathsf{R} \to \mathsf{R}$ satisfies the functional relation

$$f(x+y) = f(x)+f(y)$$

for all x, y in R. Show that $f(\lambda x) = \lambda f(x)$, for any x in R,

(i) when $\lambda = 0, 1, 2, \ldots$

(ii) when λ is any integer, and lastly

(iii) when λ is any rational.

If f is also assumed to be continuous deduce that $f(x) = cx$ for all x in R for some real constant c.

2. In the notation of Definition 7.1.1 show that the following statements are equivalent:

(i) $f(y)$ tends to x' as y tends to x in Y;

(ii) the function $g\colon(Z, \rho_Z) \to (X', \rho')$ is continuous at x, where $Z = Y \cup \{x\}$ and g is defined by $g(y) = f(y)$ if $y \in Y - \{x\}$ and $g(x) = x'$;

(iii) for every sequence (x_n) in $Y - \{x\}$ which converges to x, the sequence $(f(x_n))$ converges to x'.

3. Let (X, ρ), (X', ρ'), (X'', ρ'') be metric spaces, $Y \subseteq X$ and x be a limit point of Y. If $f:(Y, \rho_Y) \to (X', \rho')$ and $g:(X', \rho') \to (X'', \rho'')$ are functions such that $f(y)$ tends to a limit x' (in X') as y tends to x in Y, and g is continuous at x', then show that $g(f(y))$ tends to $g(x')$ as y tends to x in Y.

Give an example to show that the requirement that g is continuous at x' cannot be replaced by the weaker one that $g(y')$ tends to a limit as y' tends to x' in X'.

7.2 Two extension theorems

We now establish our first result concerning the extension of continuous functions.

THEOREM 7.2.1. *Let (X, ρ), (X', ρ') be two metric spaces and Y be everywhere dense in (X, ρ); let $f:(Y, \rho_Y) \to (X', \rho')$ be continuous. Then there exists a continuous extension $F:(X, \rho) \to (X', \rho')$ of f if and only if for each x in $X - Y$ the limit*

$$\lim_{y \to x,\, y \in Y} f(y) \tag{7.2.1}$$

exists. If such a continuous extension exists, then it is unique.

Proof. First assume that there exists a continuous function $F:(X, \rho) \to (X', \rho')$ such that $F(x) = f(x)$ for all x in Y. Then by the continuity of F, the limit

$$\lim_{y \to x,\, y \in Y} F(y)$$

exists and so (7.2.1) exists.

Conversely assume that the limit (7.2.1) exists and define a function $F:(X, \rho) \to (X', \rho')$ by $F(x) = f(x)$ if $x \in Y$, and

$$F(x) = \lim_{y \to x,\, y \in Y} f(y)$$

if $x \notin Y$; thus F is an extension of f. It remains to show that F is continuous over X.

Let $x_0 \in X$ and let $\varepsilon > 0$ be given. If $y \in Y$ then $F(y) = f(y)$, so by the definition of F, there exists $\delta > 0$ such that

$$f(Y \cap S(x_0, \delta)) \subseteq S'(F(x_0), \tfrac{1}{2}\varepsilon),$$

where S' denotes a sphere in (X', ρ'). Now let $x \in (X-Y) \cap S(x_0, \delta)$; then x is a limit point of $Y \cap S(x_0, \delta)$. Hence by Lemma 7.1.2

$$F(x) \in \overline{S'(F(x_0), \tfrac{1}{2}\varepsilon)} \subseteq S'(F(x_0), \varepsilon).$$

Therefore given any $\varepsilon > 0$ there exists $\delta > 0$ such that $\rho'(F(x), F(x_0)) < \varepsilon$ for all x such that $\rho(x, x_0) < \delta$, so F is continuous at x_0.

By the comment following Lemma 7.1.1 it follows that F must be unique.

The second result concerns the extension of uniformly continuous functions.

THEOREM 7.2.2. *Let (X, ρ), (X', ρ') be two metric spaces, Y be everywhere dense in (X, ρ) and (X', ρ') be complete; let the function $f:(Y, \rho_Y) \to (X', \rho')$ be uniformly continuous. Then there exists an extension $F:(X, \rho) \to (X', \rho')$ of f which is uniformly continuous over X.*

Proof. First it is shown that the hypotheses ensure that the limit

$$\lim_{y \to x,\, y \in Y} f(y) \tag{7.2.2}$$

exists for any x in $X-Y$. Now the limit (7.2.2) exists, and is equal to x', if and only if $y_n \to x$ (as $n \to \infty$) always implies that $f(y_n) \to x'$ where $y_n \in Y-\{x\}$ for each n (see Exercise 7.1.2). Of course, since $x \notin Y$, $Y-\{x\} = Y$.

Let (y_n) be a sequence in Y such that $y_n \to x$; then (y_n) is a fundamental sequence. Since f is uniformly continuous, it easily follows (see, for example, Exercise 4.1.5) that $(f(y_n))$ is a fundamental sequence in (X', ρ'); but (X', ρ') is complete so there exists x' in X' such that $f(y_n) \to x'$ as $n \to \infty$. Moreover if (z_n) is a second sequence in Y which converges to x then it is easily seen that $f(z_n) \to x'$. Therefore (7.2.2) exists and so there exists a continuous extension $F:(X, \rho) \to (X', \rho')$ of f. It remains to show that F is uniformly continuous.

Given any $\varepsilon > 0$ there exists $\delta > 0$ for which $\rho'(f(x), f(y)) < \varepsilon$ for all x, y in Y such that $\rho(x, y) < 3\delta$. Let p, q in X be such that $\rho(p, q) < \delta$. Then there exist infinite sequences (x_n), (y_n) in Y such that $x_n \to p$, $y_n \to q$ as $n \to \infty$; hence there exists N such that $\rho(p, x_n) < \delta$, $\rho(q, y_n) < \delta$ for all $n \geqslant N$, and so $\rho(x_n, y_n) < 3\delta$ for all $n \geqslant N$. Thus $\rho'(f(x_n), f(y_n)) < \varepsilon$, that is, $\rho'(F(x_n), F(y_n)) < \varepsilon$ for all $n \geqslant N$. Since F is continuous, using Lemma 2.3.1, on letting $n \to \infty$ it follows that $\rho'(F(p), F(q)) \leqslant \varepsilon$. Since f is uniformly continuous over Y, the original δ may be assumed to be independent of x, y in Y, so F is uniformly continuous over X.

Of course Theorem 7.2.1 tells us that for any continuous function f defined on a subset Y of a metric space (X, ρ) there is a continuous extension of f to $\bar{Y}$ provided $f(y)$ tends to a limit as $y \to x$ (in Y) for all x in $\bar{Y} - Y$; a similar comment applies to Theorem 7.2.2.

EXERCISES 7.2

1. Show that the function $f:((0, 1], \mathrm{d}) \to ([-1, 1], \mathrm{d})$ defined by $f(x) = \sin x^{-1}$ cannot be extended continuously to $[0, 1]$.

2. In the notation of Theorem 7.2.2 let $X = \mathsf{R}$, $X' = \mathsf{Q}$ and ρ, ρ' be the relevant Euclidean metrics; also let $Y = \mathsf{Q}$. Show that the function $f:(Y, \rho_Y) \to (X', \rho')$ defined by $f(x) = x$ for all x in Y is continuous but that it possesses no extension continuous over X. [Thus the requirement in Theorem 7.2.2 that (X', ρ') be complete cannot be omitted.]

7.3 The Tietze extension theorem

In this section we establish an important extension result known as the Tietze theorem. In the previous section our extensions were from sets to their closures; in the Tietze theorem the extension is of a real-valued continuous function defined on a closed proper subset of (X, ρ) to the whole of X. First a preliminary result is established.

LEMMA 7.3.1. *Let (X, ρ) be a metric space and Y be a non-empty closed subset of (X, ρ); let $f:(Y, \rho_Y) \to (\mathsf{R}, \mathrm{d})$ be a bounded continuous function for which there exists $M > 0$ such that*

$$\inf_{x \in Y} f(x) = -M, \qquad \sup_{x \in Y} f(x) = M. \tag{7.3.1}$$

Then there exists a continuous function $g:(X, \rho) \to (\mathsf{R}, \mathrm{d})$ such that $|g(x)| \leqslant \frac{1}{3}M$ for all x and

$$|g(x)| < \tfrac{1}{3}M \text{ on } X - Y, \quad |f(x) - g(x)| \leqslant \tfrac{2}{3}M \text{ on } Y.$$

Proof. Let

$$A = \{x:f(x) \leqslant -\tfrac{1}{3}M\}, \quad B = \{x:f(x) \geqslant \tfrac{1}{3}M\};$$

then the sets A, B are non-empty, disjoint and, by (ii) of Lemma 7.1.1, closed in (Y, ρ_Y). Since Y is closed in (X, ρ), A, B are closed in (X, ρ); why? Define $g:(X, \rho) \to (\mathsf{R}, \mathsf{d})$ by

$$g(x) = \tfrac{1}{3}M\frac{\rho(x, A)-\rho(x, B)}{\rho(x, A)+\rho(x, B)}.$$

Then, by Lemma 3.5.1, g is continuous.

Clearly $|g(x)| \leqslant \tfrac{1}{3}M$ for all x in X. Moreover $x \in X-Y$ implies that $x \notin A$ and $x \notin B$, so $|g(x)| < \tfrac{1}{3}M$. Lastly suppose that $x \in Y$. If $x \in A$ then $g(x) = -\tfrac{1}{3}M$, $-M \leqslant f(x) \leqslant -\tfrac{1}{3}M$ so $|f(x)-g(x)| \leqslant \tfrac{2}{3}M$; similarly for $x \in B$. If $x \in Y-(A \cup B)$ then $|g(x)| < \tfrac{1}{3}M$, $|f(x)| < \tfrac{1}{3}M$ so $|f(x)-g(x)| < \tfrac{2}{3}M$.

THEOREM 7.3.1 (The Tietze extension theorem). *Let (X, ρ) be a metric space and Y be a non-empty closed subset of (X, ρ); let $f:(Y, \rho_Y) \to (\mathsf{R}, \mathsf{d})$ be continuous. Then there exists a continuous extension $F:(X, \rho) \to (\mathsf{R}, \mathsf{d})$ of f.*

Moreover if f is bounded, say

$$m = \inf_{x \in Y} f(x), \quad M = \sup_{x \in Y} f(x),$$

then, provided $m < M$, there exists a continuous extension $F:(X, \rho) \to (\mathsf{R}, \mathsf{d})$ such that $m < F(x) < M$ on $X-Y$.

Proof. First it is assumed that f is bounded; without loss of generality we can suppose that $m = -M$ (for otherwise a constant can be added to f to achieve this).

We define a sequence (g_n) of continuous functions $g_n:(X, \rho) \to (\mathsf{R}, \mathsf{d})$ as follows. Write f_1 in place of f; by Lemma 7.3.1 there exists a continuous function $g_1:(X, \rho) \to (\mathsf{R}, \mathsf{d})$ such that $|g_1(x)| \leqslant \tfrac{1}{3}M$ for all x and

$$|g_1(x)| < \tfrac{1}{3}M \text{ on } X-Y, \quad |f_1(x)-g_1(x)| \leqslant \tfrac{2}{3}M \text{ on } Y.$$

Let $f_2:(Y, \rho_Y) \to (\mathsf{R}, \mathsf{d})$ be defined by $f_2(x) = f_1(x)-g_1(x)$; then there exists a continuous function $g_2:(X, \rho) \to (\mathsf{R}, \mathsf{d})$ such that $|g_2(x)| \leqslant \tfrac{1}{3}(\tfrac{2}{3}M)$ for all x and

$$|g_2(x)| < \tfrac{1}{3}(\tfrac{2}{3}M) \text{ on } X-Y, \quad |f_2(x)-g_2(x)| \leqslant (\tfrac{2}{3})^2 M \text{ on } Y.$$

Similarly let $f_3:(Y, \rho_Y) \to (\mathsf{R}, \mathsf{d})$ be defined by $f_3(x) = f_2(x) - g_2(x)$; then there exists a continuous function $g_3:(X, \rho) \to (\mathsf{R}, \mathsf{d})$ such that $|g_3(x)| \leqslant \frac{1}{3}(\frac{2}{3})^2 M$ for all x, and

$$|g_3(x)| < \tfrac{1}{3}(\tfrac{2}{3})^2 M \text{ on } X-Y, \quad |f_3(x)-g_3(x)| \leqslant (\tfrac{2}{3})^2 M \text{ on } Y.$$

Proceeding in this way it follows that for each positive integer n there exists a function $g_n:(X, \rho) \to (\mathsf{R}, \mathsf{d})$ such that

$$|g_n(x)| \leqslant \tfrac{1}{3}(\tfrac{2}{3})^{n-1} M \text{ for all } x, \tag{7.3.2}$$

$$|g_n(x)| < \tfrac{1}{3}(\tfrac{2}{3})^{n-1} M \text{ on } X-Y, \tag{7.3.3}$$

and

$$|f_n(x)-g_n(x)| < (\tfrac{2}{3})^n M \text{ on } Y,$$

where

$$f_n = f_{n-1}-g_{n-1} = f_{n-2}-(g_{n-2}+g_{n-1}) = \ldots = f-(g_1+\ldots+g_{n-1})$$

so

$$\left|f(x) - \sum_{i=1}^{n} g_i(x)\right| \leqslant (\tfrac{2}{3})^n M \text{ on } Y. \tag{7.3.4}$$

Observe that, as n increases, $g_1 + \ldots + g_n$ becomes a closer and closer approximation to f on Y, and that $g_1+\ldots+g_n$ is a continuous function defined on the whole of X.

Define $F:(X, \rho) \to (\mathsf{R}, \mathsf{d})$ by

$$F(x) = \sum_{i=1}^{\infty} g_i(x) \tag{7.3.5}$$

for all x in X. By (7.3.2) and the Weierstrass M-test, the series (7.3.5) is uniformly convergent over X; but each g_i is continuous so F is continuous.

Letting $n \to \infty$ in (7.3.4) it follows that $F(x) = f(x)$ on Y. Moreover, from (7.3.3) for all x in $X-Y$

$$|F(x)| \leqslant \sum_{i=1}^{\infty} |g_i(x)| < M \sum_{i=1}^{\infty} \frac{2^{i-1}}{3^i} = M,$$

and so $|F(x)| < M$ on $X-Y$.

Now suppose that f is not bounded over Y. Let h be any homeomorphism of R onto the open interval $(-1, 1)$; for example let h be defined by

$$h(x) = \left(\frac{2}{\pi}\right) \tan^{-1} x$$

for all x in R. Then the function $h \circ f$ is continuous and bounded over X; hence there exists a continuous extension $H:(X,\rho)\to(\mathsf{R},\mathsf{d})$ of h. Define $F:(X,\rho)\to(\mathsf{R},\mathsf{d})$ by $F = h^{-1}\circ H$; then F is continuous over X and $F(x) = f(x)$ on Y. This completes the proof.

In conclusion we make some remarks about Tietze's theorem.

Firstly note that if $Y \subset X$ then F is not unique; in fact there then exists an infinite number of extensions (see Exercise 7.3.1). Secondly the assumption that Y is a closed subset of (X, ρ) is essential; for example the function $f:((0, 1], \mathsf{d}) \to (\mathsf{R}, \mathsf{d})$ defined by $f(x) = \sin(1/x)$ cannot be extended continuously to $[0, 1]$.

EXERCISES 7.3

1. Let Y be a proper closed subset of a metric space (X, ρ) and $f:(Y, \rho_Y) \to (\mathsf{R}, \mathsf{d})$ be continuous. Let $x_0 \in X - Y$, $Z = Y \cup \{x_0\}$ and α be any real number. Show that the function $g:(Z, \rho_Z)\to(\mathsf{R}, \mathsf{d})$, defined by $g(x) = f(x)$ if $x \in Y$ and $g(x_0) = \alpha$, is continuous. Deduce that there exist an infinite number of continuous extensions as specified in Theorem 7.3.1.

2. State and prove a generalization of Tietze's theorem to continuous functions $f:(Y, \rho_Y) \to (\mathsf{R}^m, \mathsf{d})$.

3. Let (X, ρ) be a metric space which is not compact, so there is a sequence (x_n) of distinct points in X having no cluster point. Let $Y = \{x_i : i = 1, 2, \ldots\}$; show that the function $f:(Y, \rho_Y) \to (\mathsf{R}, \mathsf{d})$ defined by $f(x_i) = i$ is continuous over Y. Deduce that there exists a continuous function $F:(X, \rho) \to (\mathsf{R}, \mathsf{d})$ which is unbounded.

Show also that the function $g:(Y, \rho_Y) \to (\mathsf{R}, \mathsf{d})$ defined by

$$g(x_i) = (-1)^i(1-i^{-1})$$

is continuous. Deduce that there exists a continuous function $G:(X, \rho) \to (\mathsf{R}, \mathsf{d})$ which is bounded but does not attain its bounds.

Hence establish the following result.

Let (X, ρ) be a metric space; then the following statements are equivalent:

(i) (X, ρ) is compact;

(ii) every real-valued continuous function on X is bounded;

(iii) every bounded real-valued continuous function on X attains its bounds.

4. Let (Y_n) be a sequence of closed sets in a metric space (X, ρ); let Y denote the union of all the sets Y_n, and suppose that there exists x_0 in $X-Y$. With the aid of Exercise 3.1.2 prove that there exists a real-valued continuous function on X such that $f(y) > 0$ for all y in Y and $f(x_0) = 0$.

B. BAIRE'S CATEGORY THEOREM

7.4 Nowhere dense sets

In §2.10 it was explained that a subset Y of a metric space (X, ρ) is said to be everywhere dense if $\bar{Y} = X$, that is, roughly speaking, if 'Y nearly fills X out'. We now introduce subsets of the other extreme, that is, ones which cover virtually nothing of the space.

DEFINITION 7.4.1. A subset Y of a metric space (X, ρ) is said to be *nowhere dense* (*in* (X, ρ)) if $\bar{Y}$ contains no interior points, that is, if $(\bar{Y})^\circ = \varnothing$.

We make some remarks.

(i) The empty set is nowhere dense in every metric space.

(ii) In any discrete metric space $\varnothing$ is the only subset which is nowhere dense; for any non-empty subset Y is both open and closed so

$$(\bar{Y})^\circ = Y^\circ = Y \neq \varnothing.$$

(iii) If Y is nowhere dense in a metric space (X, ρ) then any subset Z of Y is also nowhere dense in (X, ρ).

(iv) In a metric space (X, ρ), for any x in X the set $\{x\}$ is nowhere dense in (X, ρ) if and only if x is not an isolated point† of (X, ρ); thus if (X, ρ) has no isolated points then $\{x\}$ is nowhere dense for every x in X.

(v) Clearly the property of a set being nowhere dense is not the negation of a set being everywhere dense. However we do have the following result connecting the two concepts.

LEMMA 7.4.1. *If (X, ρ) is a metric space and $Y \subset X$, then Y is nowhere dense if and only if $X - \bar{Y}$ is everywhere dense.*

† Isolated points were defined in §2.5.

Proof. By Lemma 2.11.1 we have

$$\bar{Y}^\circ = X - \overline{X - \bar{Y}},$$

so $\bar{Y}^\circ = \varnothing$ if and only if $\overline{X - \bar{Y}} = X$.

COROLLARY 7.4.1. *If Y is nowhere dense in (X, ρ), then $X - Y$ is everywhere dense in (X, ρ). The converse is false.*

Proof. The first part follows from Lemma 7.4.1 and $X - \bar{Y} \subseteq X - Y$. That the converse is false is shown by the following example. In (R, d), $\mathsf{R} - \mathsf{Q}$ (the set of irrationals) is everywhere dense, but $\bar{\mathsf{Q}} = \mathsf{R}$ so Q is certainly not nowhere dense. (That $\mathsf{R} - \mathsf{Q}$ is everywhere dense can be shown by an argument similar to that used in §2.10 to show that Q is everywhere dense.)

LEMMA 7.4.2. *Let (X, ρ) be a metric space and $Z \subseteq Y \subseteq X$. If Z is nowhere dense in (Y, ρ_Y) then Z is nowhere dense in (X, ρ).*

Proof. This is left to the reader.

LEMMA 7.4.3. *Let Y be a subset of a metric space (X, ρ); then the following statements are equivalent:*

(i) *Y is nowhere dense;*

(ii) *$\bar{Y}$ does not contain a non-empty open set;*

(iii) *each non-empty open set U contains a non-empty open subset V such that $V \cap \bar{Y} = \varnothing$;*

(iv) *each non-empty open set U contains a non-empty open subset V such that $V \cap Y = \varnothing$;*

(v) *each non-empty open set U contains an open sphere $S(x, \varepsilon)$ such that $S(x, \varepsilon) \cap Y = \varnothing$;*

(vi) *each non-empty open set U contains an open sphere $S(x, \varepsilon)$ such that $S(x, \varepsilon) \cap \bar{Y} = \varnothing$.*

Proof. Most of this is trivial. The equivalence of (i) and (ii) follows immediately from the definition of a nowhere dense set.

Assume (i). Then $X - \bar{Y}$ is everywhere dense; it is open and so meets any non-empty open set U in an open set V, say. Thus $V \cap \bar{Y} = \varnothing$; moreover $V \neq \varnothing$, for if $V = \varnothing$ then $(X - \bar{Y}) \cap U = \varnothing$ and so $U \subseteq \bar{Y}$, which contradicts (ii). This establishes (iii).

Trivially (iii) implies (iv) and (iv) implies (v).

Assume (v) so that $S(x, \varepsilon) \cap Y = \varnothing$; then $S(x, \frac{1}{2}\varepsilon) \cap \bar{Y} = \varnothing$ and (vi) is established.

Assume (vi) and suppose, if possible, that $(\bar{Y})^\circ \neq \varnothing$; let $S(x, \varepsilon) \subseteq \bar{Y}$ where $\varepsilon > 0$. Then there exists a sphere $S(x', \varepsilon')$ such that $S(x', \varepsilon') \subseteq S(x, \varepsilon)$ and such that $S(x', \varepsilon') \cap \bar{Y} = \varnothing$; this gives the necessary contradiction, and completes the proof.

LEMMA 7.4.4. *The union of a finite number of nowhere dense sets is nowhere dense.*

Proof. Let Y_1, Y_2 be nowhere dense in (X, ρ) and let U be an arbitrary non-empty open subset of (X, ρ). Then by Lemma 7.4.3 there exists a non-empty open set V_1 such that $V_1 \subseteq U$ and $V_1 \cap Y_1 = \varnothing$. Similarly there exists a non-empty open set V_2 such that $V_2 \subseteq V_1 \subseteq U$ and $V_2 \cap Y_2 = \varnothing$. Then

$$V_2 \cap (Y_1 \cup Y_2) = (V_2 \cap Y_1) \cup (V_2 \cap Y_2) = \varnothing;$$

thus U contains a non-empty open set V_2 such that $V_2 \cap (Y_1 \cup Y_2) = \varnothing$, so $Y_1 \cup Y_2$ is nowhere dense in (X, ρ). By induction the union of any finite number of nowhere dense sets is nowhere dense.

However the union of a countably infinite number of nowhere dense sets may, or may not, be nowhere dense. For example, in (R, d) the sets

$$\bigcup_{x \in \mathsf{Z}} \{x\}, \quad \bigcup_{x \in \mathsf{Q}} \{x\}$$

are nowhere dense and not nowhere dense, respectively.

If Y is nowhere dense in a metric space (X, ρ), then $\bar{Y} \neq X$; thus in view of Lemma 7.4.4 it follows that if $Y_1, \ldots, Y_n$ are a finite number of nowhere dense sets in (X, ρ), then their union cannot be all of X. A question which arises from this is whether or not the union of a countably infinite number of nowhere dense sets of (X, ρ) can be all of X. A partial answer to this is the following result of Baire, namely that if (X, ρ) is complete then it certainly cannot be expressed as the union of a countable number of nowhere dense sets. This result is proved in the next section.

EXERCISES 7.4

1. Give an example to show that the converse of Lemma 7.4.2 is false.

2. If the mapping $f:(X, \rho) \to (X', \rho')$ is a homeomorphism, prove that a set Y is nowhere dense in (X, ρ) if and only if $f(Y)$ is nowhere dense in (X', ρ').

3. In the notation of Exercise 2.2.2, let $Y_i \subseteq X_i$ for each i. Show that $Y_1 \times \ldots \times Y_m$ is nowhere dense in (X, ρ), or in (X', ρ'), if and only if at least one of the sets Y_i is nowhere dense in (X_i, ρ_i).

4. Let Y be a subset of the Euclidean space $(\mathsf{R}^m, \mathsf{d})$ such that Y possesses only a finite number of limit points; prove that Y is nowhere dense. Show that this result does not hold in an arbitrary metric space.

5. Prove Lemma 7.4.4 using the definition of a nowhere dense set instead of appealing to Lemma 7.4.3.

6. Show that the boundary (see §2.11) of a set Y is nowhere dense in a metric space if Y is either open or closed.

7.5 Categories and Baire's category theorem

DEFINITION 7.5.1. Let (X, ρ) be a metric space and $Y \subseteq X$; then Y is said to be a set of the *first category*, or *meagre*, *in* (X, ρ) if it can be represented as the union of a countable number of nowhere dense sets of (X, ρ); otherwise it is said to be of the *second category*, or *non-meagre*, *in* (X, ρ).

The terms meagre and non-meagre are to be preferred to the less descriptive ones of the first and second category which were originally introduced by Baire; however these new terms are not, as yet, in general usage.

We now give some examples.

(i) In (R, d), the set Q is meagre since every singleton set $\{q\}$, where $q \in \mathsf{Q}$, is nowhere dense.

(ii) In any discrete metric space the only set which is nowhere dense is the empty set, so every non-empty subset of a discrete space is non-meagre.

(iii) In $(\mathsf{R}^2, \mathsf{d})$ any finite set of points is meagre; the set $\{(x, 0): x \in \mathsf{R}\}$ is also meagre (in fact it is nowhere dense).

If $Z \subseteq Y \subseteq X$ in a metric space (X, ρ) and Y is meagre in (X, ρ), then Z must also be meagre in (X, ρ); on the other hand, if Z is non-meagre in (X, ρ) then Y will be non-meagre in (X, ρ). The verification of these assertions together with that of the next result (which is the extension of Lemma 7.4.2) is left to the reader.

LEMMA 7.5.1. *Let* (X, ρ) *be a metric space and* $Z \subseteq Y \subseteq X$.

(i) *If* Z *is meagre in* (Y, ρ_Y) *then* Z *is meagre in* (X, ρ).

(ii) *If* Z *is non-meagre in* (X, ρ) *then* Z *is non-meagre in* (Y, ρ_Y).

LEMMA 7.5.2. *The union of a sequence of meagre sets (in a common metric space) is also a meagre set.*

Proof. Let (Y_n) be a sequence of meagre sets; then each Y_n can be expressed as the union of a countable number of nowhere dense sets; but the union of a countable number of countable sets is also countable, so the required result follows.

We now come to a basic result, namely Baire's category theorem, which asserts that any complete metric space is non-meagre, that is, it is a set of the second category (in itself). The essence of this result is contained in the next theorem whose proof depends on Cantor's intersection theorem (see §4.3).

THEOREM 7.5.1. *If* (Y_n) *is a sequence of nowhere dense sets in a complete metric space* (X, ρ), *then there exists* x *in* X *such that* $x \notin Y_n$ *for all* n.

Proof. By Lemma 7.4.3 there exists an open sphere $S(x_1, \varepsilon_1)$, where $\varepsilon_1 < 1$, such that

$$S(x_1, \varepsilon_1) \cap Y_1 = \varnothing.$$

Then $(\bar{S}(x_1, \frac{1}{2}\varepsilon_1))^\circ$ is a non-empty open set, so contains an open sphere $S(x_2, \varepsilon_2)$, where $\varepsilon_2 < \frac{1}{2}$, such that

$$S(x_2, \varepsilon_2) \cap Y_2 = \varnothing.$$

Similarly $(\bar{S}(x_2, \frac{1}{2}\varepsilon_2))^\circ$ is a non-empty open set, so contains an open sphere $S(x_3, \varepsilon_3)$ where $\varepsilon_3 < \frac{1}{3}$, such that

$$S(x_3, \varepsilon_3) \cap Y_3 = \varnothing.$$

Proceeding in this way, that is by induction, we obtain a nested sequence $(\bar{S}(x_n, \frac{1}{2}\varepsilon_n))$ of closed spheres whose radii tend to zero as $n \to \infty$. Since (X, ρ) is complete, by Cantor's intersection theorem their intersection contains a unique point x, say; but $x \in \bar{S}(x_n, \frac{1}{2}\varepsilon_n)$

and

$$\bar{S}(x_n, \tfrac{1}{2}\varepsilon_n) \cap Y_n = \emptyset,$$

so $x \notin Y_n$, for all n.

COROLLARY 7.5.1. *If (Y_n) is a sequence of nowhere dense sets in a complete metric space (X, ρ) and Z is any non-empty open subset of (X, ρ), then there exists x in Z such that $x \notin Y_n$ for all n.*

Proof. This result follows from the argument used to establish the main result with the sole modification that the first sphere $S(x_1, \varepsilon_1)$ is taken to be a subset of Z; then the point x constructed in the above proof must lie in Z.

Clearly Theorem 7.5.1 may be restated in either of the following forms.

THEOREM 7.5.1′. *If a complete metric space is the union of a sequence of its subsets, then at least one of these subsets is not nowhere dense.*

THEOREM 7.5.1″ (Baire's category theorem). *Any complete metric space is of the second category (in itself).*

Corollary 7.5.1 above implies, of course, that any non-empty open subset of a complete metric space is a set of the second category.

THEOREM 7.5.2. *If (Y_n) is a sequence of everywhere dense open sets in a complete metric space (X, ρ) then*

$$Y = \bigcap_{n=1}^{\infty} Y_n$$

is everywhere dense.

Proof. Let $Z_n = X - Y_n$; then Z_n is closed and so is nowhere dense. Let S be any open sphere in (X, ρ); then $S \cap Z_n$ is nowhere dense for all n so, by Corollary 7.5.1,

$$\bigcup_{n=1}^{\infty} (S \cap Z_n) \neq S.$$

Hence

$$S \cap \left(\bigcup_{n=1}^{\infty} Z_n \right) \neq S,$$

$$S \cap \left(X - \bigcup_{n=1}^{\infty} Z_n \right) \neq \emptyset,$$

so
$$S \cap Y \neq \varnothing.$$

Therefore Y is everywhere dense (using the elementary result that a set V is everywhere dense in a metric space (X, ρ) if and only if $S \cap V \neq \varnothing$ for every open sphere S of (X, ρ)—see Exercise 2.10.1).

COROLLARY 7.5.2. *Let* (X, ρ) *be a complete metric space and* $Y \subset X$. *If* Y *is meagre then* $X - Y$ *is everywhere dense.*

Proof. Let (Z_n) be a sequence of nowhere dense sets whose union is Y. Then $X - \bar{Z}_n$ is everywhere dense and open for all n so

$$\bigcap_{n=1}^{\infty} (X - \bar{Z}_n),$$

that is,
$$X - \bigcup_{n=1}^{\infty} \bar{Z}_n$$

is everywhere dense; but

$$X - \bigcup_{n=1}^{\infty} \bar{Z}_n \subseteq X - \bigcup_{n=1}^{\infty} Z_n = X - Y,$$

so $X - Y$ is everywhere dense.

Baire's category theorem is a powerful result in the study of complete metric spaces. In particular it is required in order to prove two of the three basic results of that part of functional analysis concerned with Banach spaces (the results are the uniform boundedness theorem and the open mapping theorem); the study of these is outside the realm of this text. However we shall illustrate the power of Baire's theorem by proving (in the next section) a result of classical analysis, namely that there exist functions $f: \mathsf{R} \to \mathsf{R}$ which are continuous at all points of R but differentiable at none.

We conclude this section with a simple application of Baire's theorem. In this application the metric is d throughout.

Let $f: [0, 1] \to \mathsf{R}$ be continuous over $[0, 1]$; then f possesses an integral f_1 (which is both a definite and an indefinite integral) over $[0, 1]$ which in turn is differentiable, and hence continuous, over $[0, 1]$. Let f_2 be any integral of f_1 and so on. If, for some n in N, $f_n(x) = 0$ for all x in $[0, 1]$ then by differentiating n times it follows that $f(x) = 0$ for all x in $[0, 1]$; we now generalize this result. Suppose, instead of being given that there exists n in N for which $f_n(x) = 0$ for all x in $[0, 1]$, it is given that for each x in $[0, 1]$ there exists n, dependent possibly on x, such that $f_n(x) = 0$. Then again it is shown that $f(x) = 0$ for all x in $[0, 1]$, but now it is necessary to use Baire's theorem.

For each positive integer n, let Y_n denote the set of all x in $[0, 1]$ such that $f_n(x) = 0$. By Lemma 7.1.1 Y_n is closed; by hypothesis

$$\bigcup_{n=1}^{\infty} Y_n = [0, 1].$$

Let n be any positive integer such that Y_n is not nowhere dense so $Y_n^\circ \neq \varnothing$; let $x_0 \in Y_n^\circ$ and ε (> 0) be such that

$$[x_0-\varepsilon, x_0+\varepsilon] \subseteq Y_n^\circ.$$

As before, $f(x) = 0$ on $[x_0-\varepsilon, x_0+\varepsilon]$; in particular $f(x_0) = 0$ and hence $f(x) = 0$ for all x in Y_n. Let $Z_n = Y_n - Y_n^\circ$; then Z_n is closed (since

$$Z_n = Y_n \cap ([0, 1] - Y_n^\circ),$$

Z_n is the intersection of two closed sets). Hence $(\bar{Z}_n)^\circ = Z_n^\circ = \varnothing$, so Z_n is nowhere dense; thus $f(x) = 0$ for all x in $[0, 1]$ except possibly on a set of the first category. Since $[0, 1]$, with Euclidean metric, is complete, by Baire's theorem it follows that $f(x)$ is zero on a subset of $[0, 1]$ of the second category, the complement of which is a set of the first category. It is left to the reader to show that, since f is continuous, $f(x) = 0$ everywhere on $[0, 1]$.

EXERCISES 7.5

1. If (X, ρ) is a complete metric space having no isolated points, prove that X must be uncountable.

2. Let (X, ρ) be of the second category. If (Y_n) is a sequence of open sets each of which is everywhere dense and if Y denotes their intersection, what is the category of the sets Y and $X - Y$?

3. In the notation of Exercise 2.2.2, let $Y_i \subseteq X_i$ for each i. Show that $Y_1 \times \ldots \times Y_m$ is meagre in (X, ρ), or in (X', ρ'), if at least one of the sets Y_i is meagre in (X_i, ρ_i).

4. Deduce Theorem 7.5.1″ from Theorem 7.5.2.

5. By mimicking the proof of Theorem 7.5.1, give a proof of Theorem 7.5.2 which is independent of Theorem 7.5.1 (or any of its equivalent forms).

6. Let (X, ρ) be a complete metric space and (f_n) be a sequence of real-valued continuous functions defined on (X, ρ) such that $(f_n(x))$ is bounded for each x in X. Show that the set

$$Y_m = \{x : |f_n(x)| \leqslant m;\ n = 1, 2, \ldots\}$$

is closed for each positive integer m. Deduce that there exists a non-empty open set Y such that $(f_n(x))$ is uniformly bounded over Y. [This result is known as Osgood's theorem; it is the central step in the proof of the uniform boundedness theorem.]

7. Let (X, ρ) be a complete metric space and let (Y_n) be a sequence of closed sets which cover (X, ρ). Prove that the union of all the sets Y_n° is everywhere dense.

7.6 The existence of everywhere continuous nowhere differentiable functions

First we make a general comment. Since the empty set is always nowhere dense, it is meagre, and therefore a non-meagre set cannot be empty; hence if we can show that a set of elements with a certain property is non-meagre (in some metric space), it follows that there must exist elements with the prescribed property. Thus Baire's theorem provides a method of establishing certain existence results; this is illustrated below.

We now show that there exist functions which are continuous over an interval $I = [a, b]$ but which are not differentiable at any point of I; moreover it will be shown that most continuous functions have this property *in the sense that* the set of all functions in $\mathscr{C}(I)$ which are differentiable at one or more points of I is meagre, when we associate the usual supremum metric ρ with $\mathscr{C}(I)$.

A function $f: I \to \mathsf{R}$ is said to have a *right-hand derivative* at c, where $a < c < b$, if there exists l in R such that, given any $\varepsilon > 0$, there exists $\delta > 0$ for which

$$\left|\frac{f(c+h)-f(c)}{h} - l\right| < \varepsilon$$

for all h such that $0 < h < \delta$. Let A denote the subset of $\mathscr{C}(I)$ consisting of all those functions which possess a right-hand derivative at one or more points of $[a, b)$. For each positive integer n let E_n

denote the set of all f in $\mathscr{C}(I)$ such that for one or more x in $[a, b-1/n]$

$$\left|\frac{f(x+h)-f(x)}{h}\right| \leqslant n$$

whenever $0 < h < 1/n$; thus $E_n \subseteq E_{n+1}$. If $f \in A$, then $f \in E_n$ for all n sufficiently large so that

$$A \subseteq \bigcup_{n=1}^{\infty} E_n.$$

It will be shown that E_n is nowhere dense for all n, so A is meagre. Similarly if B denotes the subset of $\mathscr{C}(I)$ consisting of all those functions which possess a left-hand derivative at one or more points of $(a, b]$, then B is also meagre, so $A \cup B$ is meagre. Since $(\mathscr{C}(I), \rho)$ is complete, by Baire's theorem $\mathscr{C}(I)-(A \cup B)$ is non-meagre and so is non-empty; thus there exist functions which do not possess a left- or right-hand derivative at any point of $[a, b]$, and the set of such functions is non-meagre.

It remains to prove that E_n is nowhere dense for all n.

First we show that E_n is closed in $(\mathscr{C}(I), \rho)$ by proving that its complement F_n is open. Let n be fixed and let $f \in F_n$, so for each x in $[a, b-1/n]$ there exists h_x such that $0 < h_x < 1/n$ and

$$\left|\frac{f(x+h_x)-f(x)}{h_x}\right| > n$$

(using the principles of §1.7). It has to be shown that there exists $\varepsilon > 0$ such that if $g \in \mathscr{C}(I)$ and $\rho(f, g) < \varepsilon$ then $g \in F_n$. For each x in $[a, b-1/n]$ let ε_x (> 0) be defined by

$$|f(x+h_x)-f(x)| = |h_x|n+4\varepsilon_x;$$

since f is continuous at x there exists $\delta_x > 0$ such that

$$|f(y+h_x)-f(y)| > |h_x|n+2\varepsilon_x$$

for all y in $(x-\delta_x, x+\delta_x) \cap [a, b-1/n]$.

We now use the compactness of $[a, b-1/n]$; since the collection

$$\{(x-\delta_x, x+\delta_x) : x \in [a, b-1/n]\}$$

is an open covering of $[a, b-1/n]$, a finite number of these intervals covers $[a, b-1/n]$. Let

$$\{(x_i-\delta_{x_i}, x_i+\delta_{x_i}) : i = 1, \ldots, m\}$$

be such a finite covering; also let $\varepsilon = \min(\varepsilon_{x_1}, \ldots, \varepsilon_{x_m})$. Then

$$|f(y+h_{x_i})-f(y)| > |h_{x_i}|n+2\varepsilon$$

for all y in $(x_i-\delta_{x_i}, x_i+\delta_{x_i}) \cap [a, b-1/n]$ and $i = 1, \ldots, m$. Now let g be such that $\rho(f, g) < \varepsilon$ so

$$f(y)-\varepsilon < g(y) < f(y)+\varepsilon$$

for all y in I. Then for any y in $[a, b-1/n]$, and for a suitable choice of i,

$$|g(y+h_{x_i})-g(y)| > |f(y+h_{x_i})-f(y)|-2\varepsilon > |h_{x_i}|n.$$

Hence $g \in F_n$ and so F_n is open.

The next stage is to show that E_n is nowhere dense; since E_n is closed, by Lemma 7.4.1 it is sufficient to show that $\mathscr{C}(I)-E_n$ is everywhere dense. Let $f \in \mathscr{C}(I)$; since f is uniformly continuous over I, given any $\varepsilon > 0$, there exists a positive integer k for which $|f(x)-f(x')| < \frac{1}{2}\varepsilon$ for all x, x' such that $|x-x'| \leqslant (b-a)/k$, where k is independent of x, x'. Let

$$a = x_0 < x_1 < \ldots < x_k = b$$

be the partition P of I which divides I into k equal intervals. Consider the rectangle formed by the lines $x = x_{r-1}$, $x = x_r$ and $y = f(x_{r-1}) \pm \frac{1}{2}\varepsilon$. Join the points $(x_{r-1}, f(x_{r-1}))$, $(x_r, f(x_r))$ by a polygonal 'saw-tooth' curve which remains inside the rectangle and whose line-segments have slopes exceeding n in absolute value.

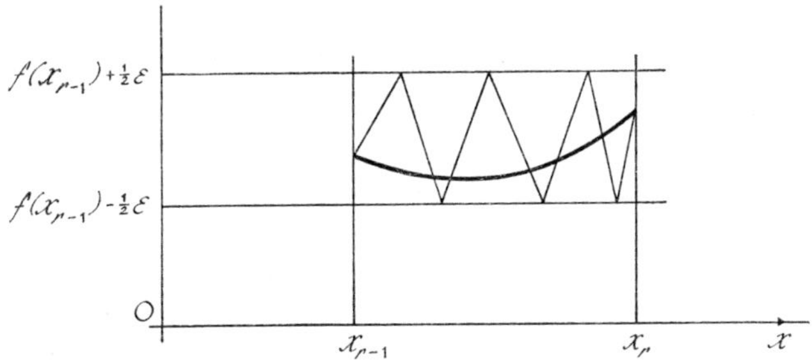

Doing this for each subinterval $[x_{r-1}, x_r]$ of P we have defined a function f_ε in $\mathscr{C}(I)$ such that $|f(x)-f_\varepsilon(x)| < \varepsilon$ for all x in I, that is, such that $\rho(f, f_\varepsilon) < \varepsilon$. Moreover f_ε is in $\mathscr{C}(I)-E_n$, so the latter is everywhere dense.

This concludes the proof.

Of course another way of proving that there exist functions which are continuous over I but are differentiable nowhere in the interval, is to find explicitly a function with these properties. This can be done although, as is only to be expected, such a function cannot be described by a simple formula; the details are outlined in Exercise 7.6.2.

Another result which is similar in spirit to that above is the following. There exist functions which are continuous on an interval I but which are not monotonic over any subinterval of I; as before most continuous functions have this property *in the sense that* the set of functions in $\mathscr{C}(I)$ which are monotonic over some subinterval of I is meagre.† For details of this see Exercise 7.6.3 and, for the actual construction of such a function see, again, Exercise 7.6.2.

EXERCISES 7.6

1. Let $I = [a, b]$. Show that there exists f in $\mathscr{C}(I)$ which is differentiable at no point of I and such that $f(a) = f(b)$. Deduce that there exists a function $f: \mathsf{R} \to \mathsf{R}$ which is continuous everywhere and is differentiable nowhere.

2. (i) Let H be the set of all rationals q of the form $q = k/4^m$ where $k \in \mathsf{Z}$, $m \in \mathsf{N}$; show that $\overline{H} = \mathsf{R}$.

(ii) Let $f_1: \mathsf{R} \to \mathsf{R}$ be defined by $f_1(x) = |x|$ if $|x| \leqslant \frac{1}{2}$ and defined periodically with period 1 for all other values of x. For any integer $n > 1$, let $f_n: \mathsf{R} \to \mathsf{R}$ be defined by

$$f_n(x) = \frac{f_1(4^{n-1}x)}{4^{n-1}}.$$

Then f_n has period $1/4^{n-1}$ and $|f_n(x)| \leqslant 1/(2.4^{n-1})$ for all x. Lastly let $f: \mathsf{R} \to \mathsf{R}$ be defined by

$$f(x) = \sum_{n=1}^{\infty} f_n(x).$$

Show that this series is uniformly convergent over R so f is continuous.

† The reader who is familiar with the result (of Lebesgue theory) 'a function which is monotonic on an interval I possesses a derivative almost everywhere in I' will see that the present result can be deduced immediately from the previous one concerning the existence of continuous functions which are nowhere differentiable.

(iii) Let $k \in \mathsf{Z}$, $m \in \mathsf{N}$, $q = k/4^m$, $h_m = 1/4^{2m+1}$. Show that

$$f_n(q) = 0 \text{ if } n > m,$$

$$f_n(q \pm h_m) = 0 \text{ if } n > 2m+1,$$

$$f_n(q \pm h_m) - f_n(q) \geqslant -h_m \text{ if } n = 1, \ldots, m,$$

$$f_n(q \pm h_m) = h_m \text{ if } n = m+1, \ldots, 2m+1.$$

Deduce that $f(q \pm h_m) - f(q) > 0$.

Using (i) show that f is monotonic on no interval in R.

(iv) Let $x \in \mathsf{R}$ and let $k_m = \pm 1/4^m$; show that for at least one of these values of k_m

$$|f_m(x+k_m) - f_m(x)| = |k_m|.$$

For such a value of k_m show also that

$$|f_n(x+k_m) - f_n(x)| = \begin{cases} |k_m| & \text{if } n \leqslant m \\ 0 & \text{if } n > m. \end{cases}$$

Hence show that

$$\frac{f(x+k_m) - f(x)}{k_m}$$

is an integer which is even if m is even and is odd if m is odd. Deduce that f cannot be differentiable at x.

3. (i) Let $I = [a, b]$. Show that the set of all subintervals of I with rational endpoints is countable. Denote these intervals by $\{I_n : n \in \mathsf{N}\}$.

(ii) Associate with $\mathscr{C}(I)$ the usual supremum metric ρ. Let E_n denote the set of all elements of $\mathscr{C}(I)$ which are monotonic on I_n; let $F_n = \mathscr{C}(I) - E_n$. Then $f \in F_n$ if and only if there exist x, y, z in I_n such that $x < y < z$ and either $f(x) < f(y)$ and $f(z) < f(y)$ or $f(x) > f(y)$ and $f(z) > f(y)$. If $f \in F_n$ show that there exists $\varepsilon > 0$ such that if $g \in S(f, \varepsilon)$ then $g \in F_n$. Deduce that E_n is a closed subset of $(\mathscr{C}(I), \rho)$.

(iii) Prove that F_n is everywhere dense in $\mathscr{C}(I)$.

Hence show that the set of functions which are continuous over I and monotonic over some sub-interval of I is meagre.

4. Let $f_n : \mathsf{R} \to \mathsf{R}$ be continuous for each n in N, and suppose that $f_n(x) \to f(x)$ as $n \to \infty$ for all x in R. Let

$$U_{mn} = \{x : |f_n(x) - f_k(x)| \leqslant m^{-1} \text{ for all } k \geqslant n\}.$$

Show that, for each m,

$$\mathsf{R} = \bigcup_{n=1}^{\infty} U_{mn}.$$

Show further that U_{mn} is closed and hence that $U_{mn} - U^{\circ}_{mn}$ is closed and nowhere dense; thus the set

$$V = \bigcup_{m,n=1}^{\infty} (U_{mn} - U_{mn})$$

is meagre.

Finally prove that f is continuous at any point not in V.

[The reader will recall that if $f_n(x) \to f(x)$ uniformly over R then f must be continuous over R. Thus by the present result even when the convergence is not uniform, the set of points of discontinuity of f is meagre.]

C. APPROXIMATION TO CONTINUOUS FUNCTIONS

7.7 The Weierstrass approximation theorem

In this section we prove two simple results concerning approximation to real-valued continuous functions and then establish the important result known as the Weierstrass approximation theorem. Although the latter is not strictly part of the general theory of metric spaces, it is included here in preparation for a brief discussion of the more general result of Stone. All three approximation results of this section depend essentially on the fact that a function which is continuous over a closed bounded set of R is uniformly continuous there. As usual I denotes the interval $[a, b]$.

DEFINITION 7.7.1. Let $g:I \to \mathsf{R}$; then g is called a *step function* if it assumes only a finite number of distinct values in R, each value being taken on some interval of R.

THEOREM 7.7.1. *Let* $f:I \to \mathsf{R}$ *be continuous; then given any* $\varepsilon > 0$ *there exists a step function* $g:I \to \mathsf{R}$ *such that* $|f(x)-g(x)| < \varepsilon$ *for all* x *in* I.

Proof. Let $\varepsilon > 0$ be given. Since f is uniformly continuous over I there exists $\delta > 0$ for which $|f(x)-f(y)| < \varepsilon$ for all x, y in I such that $|x-y| < \delta$, where δ is independent of x, y. Let P be any partition

$$a = x_0 < x_1 < \ldots < x_n = b$$

of I such that $\|P\| < \delta$; let $y_i \in (x_i, x_{i+1})$ for $i = 0, 1, \ldots, n-1$. Define a function $g: I \to \mathsf{R}$ by $g(x) = f(y_i)$ for $x_i \leqslant x < x_{i+1}$ $(i = 0, 1, \ldots, n-2)$ and $g(x) = f(y_{n-1})$ for $x_{n-1} \leqslant x \leqslant x_n$. Then it is clear that g has the required properties.

DEFINITION 7.7.2. Let $g: I \to \mathsf{R}$; then g is said to be a *polygonal* function if there exists a partition

$$a = c_0 < c_1 < \ldots < c_n = b$$

of I, and real numbers A_i, B_i, $i = 0, 1, \ldots, n-1$ such that

$$g(x) = A_i x + B_i \quad \text{for } c_i \leqslant x \leqslant c_{i+1},$$

for $i = 0, 1, \ldots, n-1$, and such that

$$A_i c_{i+1} + B_i = A_{i+1} c_{i+1} + B_{i+1}$$

for $i = 0, 1, \ldots, n-2$.

THEOREM 7.7.2. *Let $f: I \to \mathsf{R}$ be continuous; then given any $\varepsilon > 0$ there exists a polygonal function $g: I \to \mathsf{R}$ such that $|f(x) - g(x)| < \varepsilon$ for all x in I.*

Proof. Let P be the partition defined in the proof of Theorem 7.7.1; let g be the polygonal function obtained by joining the points $(x_i, f(x_i))$ linearly. If $x_i < x < x_{i+1}$, and

$$\theta = \frac{x - x_i}{x_{i+1} - x_i},$$

so $0 < \theta < 1$, then

$$\begin{aligned} g(x) &= g(x_i) + \theta\{g(x_{i+1}) - g(x_i)\} \\ &= f(x_i) + \theta\{f(x_{i+1}) - f(x_i)\}. \end{aligned}$$

Therefore

$$\begin{aligned} f(x) - g(x) &= f(x) - f(x_i) - \theta\{f(x_{i+1}) - f(x_i)\} \\ &= \{f(x) - f(x_i)\}(1-\theta) + \{f(x) - f(x_{i+1})\}\theta, \end{aligned}$$

so

$$\begin{aligned} |f(x) - g(x)| &\leqslant |f(x) - f(x_i)|(1-\theta) + |f(x) - f(x_{i+1})|\theta \\ &< \varepsilon(1-\theta) + \varepsilon\theta = \varepsilon. \end{aligned}$$

Thus $|f(x) - g(x)| < \varepsilon$ for all x in I.

We come now to the main result, due to Weierstrass; there are many proofs of it, some of which give explicit expressions for the polygonal approximations. The method which we describe (due to Lebesgue) does not do this; however it has the advantage that it can be extended to a more general situation (see §7.8). We break the proof up into stages.

LEMMA 7.7.1. *The binomial expansion (in ascending powers of* x*) of* $(1-x)^{\frac{1}{2}}$ *is uniformly convergent over* $[0, 1]$.

Proof. The binomial expansion of $(1-x)^{\frac{1}{2}}$ is

$$1-\tfrac{1}{2}\left\{x+\frac{1}{2}\cdot\frac{x^2}{2!}+\ldots+\frac{1.3.\ldots.(2r-3)}{2^{r-1}}\cdot\frac{x^r}{r!}+\ldots\right\} = 1-\tfrac{1}{2}\sum_{r=1}^{\infty} a_r(x),$$

say. Now

$$0 \leqslant a_r(x) \leqslant \frac{1.3.\ldots.(2r-3)}{2^{r-1}}\cdot\frac{1}{r!} = b_r \text{ (say)}$$

for all x in $[0,1]$; then

$$r\left\{\frac{b_r}{b_{r+1}}-1\right\} = \frac{3r}{2r-1} \to \frac{3}{2}$$

as $r \to \infty$, so by Raabe's test† (see, for example, Ferrar (1938) p. 30) $\sum b_r$ is convergent. Therefore, by the Weierstrass M-test, the series $\sum a_r(x)$ is uniformly convergent over $[0, 1]$ and so the result follows.

From Lemma 7.7.1 we deduce the Weierstrass approximation theorem for the special case of $f(x) = |x|$.

LEMMA 7.7.2. *Given any* $\varepsilon > 0$ *there exists a polynomial* p *such that* $||x|-p(x)| < \varepsilon$ *for all* x *in* $[-1, 1]$.

Proof. Taking the positive root

$$x = \sqrt{\{1-(1-x^2)\}} = \sqrt{(1-t)}$$

where $t = 1-x^2$, so that $0 \leqslant t \leqslant 1$ for all x in $[-1, 1]$. Let $P_n(t)$ denote the sum of the first $n+1$ terms of the expansion of $\sqrt{(1-t)}$ in powers of t, so P_n is a polynomial of degree n; then given any $\varepsilon > 0$ there exists N such that

$$|(1-t)-P_n(t)| < \varepsilon$$

† Raabe's test is an extension of the ratio test and is sometimes useful when, as in the present case, the ratio test is inconclusive.

for all $n \geqslant N$ where N is independent of t over $[0, 1]$. Set $p(x) = P_N(1-x^2)$; then $\big||x|-p(x)\big| < \varepsilon$ for all x in $[-1, 1]$.

LEMMA 7.7.3. *Let $f: I \to \mathsf{R}$ be defined by*

$$f(x) = \begin{cases} 0 & \text{if } a \leqslant x \leqslant c \\ m(x-c) & \text{if } c < x \leqslant b, \end{cases}$$

where $a < c < b$. Then given any $\varepsilon > 0$ there exists a polynomial p such that $|f(x)-p(x)| < \varepsilon$ for all x in I.

Proof. Clearly

$$f(x) = \tfrac{1}{2}m(x-c)+\tfrac{1}{2}m|x-c|;$$

by Lemma 7.7.2 there exists a polynomial q such that

$$\big||x-c|-q(x)\big| < \varepsilon/\tfrac{1}{2}m$$

for all x in I, so there exists a polynomial p such that $|f(x)-p(x)| < \varepsilon$ for all x in I.

We can now deduce Weierstrass' approximation theorem in its general form.

THEOREM 7.7.3. *Let $f \in \mathscr{C}(I)$; then given any $\varepsilon > 0$ there exists a polynomial p such that $|f(x)-p(x)| < \varepsilon$ for all x in I.*

Proof. Given any $\varepsilon > 0$, by Theorem 7.7.2, there exists a partition P

$$a = x_0 < x_1 < \ldots < x_n = b$$

of I and a polygonal function $G: I \to \mathsf{R}$ such that $|f(x)-G(x)| < \frac{1}{2}\varepsilon$ for all x in I, where the vertices of the polygonal arc occur at the points $(x_r, f(x_r))$, $r = 0, 1, \ldots, n$.

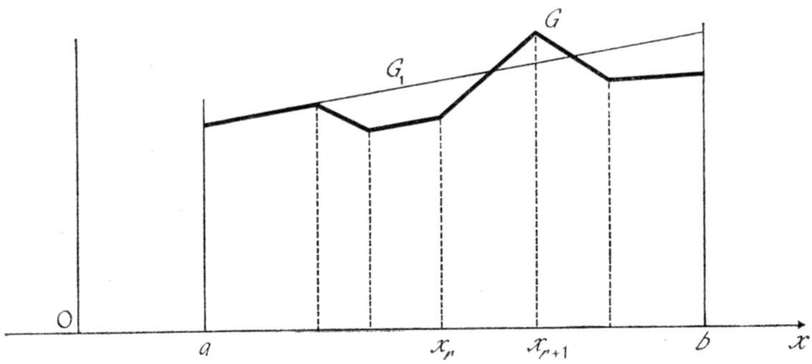

Clearly G can be expressed as the sum of n continuous functions $G_1, \ldots, G_n$ such that G_1 is linear over $[a, b]$, G_2 is zero on $[a, x_1]$ and is linear over $[x_1, b]$, G_3 is zero on $[a, x_2]$ and is linear over $[x_2, b]$, and, generally, G_r is zero on $[a, x_{r-1}]$ and is linear over $[x_{r-1}, b]$. By Lemma 7.7.3 there exist polynomials $p_1, \ldots, p_n$ such that

$$|G_r(x)-p_r(x)| < \varepsilon/2n$$

for all x in I, $r = 1, \ldots, n$. Set

$$p(x) = \sum_{r=1}^{n} p_r(x);$$

then

$$\left|G(x)-\sum_{r=1}^{n} p_r(x)\right| \leqslant \sum_{r=1}^{n} |G_r(x)-p_r(x)| < \tfrac{1}{2}\varepsilon,$$

so $|f(x)-p(x)| < \varepsilon$ for all x in I.

This concludes the proof.

Let $I_0 = [0, 1]$; then by the Weierstrass theorem the set of all finite real linear combinations of the functions

$$1, x, x^2, \ldots, x^n, \ldots$$

is dense in $\mathscr{C}(I_0)$, where the usual supremum metric is associated with $\mathscr{C}(I_0)$. There is a generalization of this result (due to Müntz) which, although we cannot prove it here, is sufficiently interesting to quote.

THEOREM 7.7.4. *The set of all finite real linear combinations of the functions*

$$1, x^{n_1}, x^{n_2}, \ldots, x^{n_r}, \ldots,$$

where (n_r) *is a strictly increasing sequence of positive integers, is dense in* $\mathscr{C}(I_0)$ *if and only if* $\sum_r n_r^{-1}$ *is divergent.*

We conclude this section with a simple application of Weierstrass' theorem.

DEFINITION 7.7.3. Let $f \in \mathscr{C}(I)$; then the members of the set

$$\{\textstyle\int_a^b x^n f(x)\,dx : n = 0, 1, 2, \ldots\}$$

are called the *moments* of f.

It is shown that any function in $\mathscr{C}(I)$ is uniquely determined by its moments.

THEOREM 7.7.5. *If* $f \in \mathscr{C}(I)$ *and*

$$\int_a^b x^n f(x)\,dx = 0 \tag{7.7.1}$$

for $n = 0, 1, 2, \ldots$, *then* $f(x) = 0$ *for all* x *in* I.

Proof. Let M be such that $|f(x)| \leqslant M$ for all x in I. By Weierstrass' theorem given any $\varepsilon > 0$ there exists a polynomial p such that

$$p(x) - \frac{\varepsilon}{M(b-a)} < f(x) < p(x) + \frac{\varepsilon}{M(b-a)},$$

for all x in I. Hence, for a given x in I, if $f(x) \geqslant 0$

$$\{f(x)\}^2 \leqslant p(x).f(x) + \frac{\varepsilon.f(x)}{M(b-a)} \leqslant p(x).f(x) + \frac{\varepsilon}{b-a},$$

and if $f(x) < 0$

$$\{f(x)\}^2 \leqslant p(x).f(x) - \frac{\varepsilon.f(x)}{M(b-a)} \leqslant p(x).f(x) + \frac{\varepsilon}{b-a}.$$

Therefore

$$\int_a^b \{f(x)\}^2\,dx \leqslant \int_a^b p(x).f(x)\,dx + \varepsilon = \varepsilon,$$

by (7.7.1); but ε is arbitrary and positive, so

$$\int_a^b \{f(x)\}^2\,dx = 0.$$

Since f is continuous, $f(x) = 0$ for all x in I.

It follows that if $f, g \in \mathscr{C}(I)$ and

$$\int_a^b x^n f(x)\,dx = \int_a^b x^n g(x)\,dx$$

for $n = 0, 1, 2, \ldots$, then $f = g$.

EXERCISES 7.7

1. Prove that

$$\left\{\frac{1.3.\ldots.(2r-1)}{r!\,2^r}\right\}^2 < \frac{1}{2r+1}$$

for $r = 1, 2, \ldots$. Deduce that the series $\sum b_r$, defined in Lemma 7.7.1, is convergent.

2. If $f: I \to \mathrm{R}$ is continuous and $f(a) = f(b) = 0$ show that, given any $\varepsilon > 0$, there exists a polynomial p such that $p(a) = p(b) = 0$ and $|f(x) - p(x)| < \varepsilon$ for all x in I.

3. A sequence (p_n) of real-valued functions is said to be *decreasing* on I if $p_n(x) \geqslant p_{n+1}(x)$ for all x in I. If $f:I \to \mathsf{R}$ is continuous prove that there exists a polynomial p_n such that

$$\left|f(x)+\frac{1}{n}-p_n(x)\right| < \frac{1}{2(n+1)^2}$$

for all x in I. Deduce that f may be approximated uniformly over I by a decreasing sequence of polynomials.

4. Let X be any compact subset of the real line. Deduce from the Weierstrass theorem (but not from Stone's generalization which is described in §7.8) that if $f: X \to \mathsf{R}$ is continuous, then, given any $\varepsilon > 0$ there exists a polynomial p such that $|f(x)-p(x)| < \varepsilon$ for all x in X.

5. Let (Y_n) be a sequence of closed subsets of R such that $Y_n \subseteq Y_{n+1}$ for $n = 1, 2, \ldots$, and Y denote the union of all the sets Y_n. Let $f: Y \to \mathsf{R}$ be a continuous function, and (ε_n) be a decreasing sequence in R such that $\varepsilon_n \to 0$ as $n \to \infty$. If (p_n) is a sequence of polynomials such that $|f(x)-p_n(x)| < \varepsilon_n$ for all x in Y_n, show that (p_n) converges to f for all x in Y.

Hence prove that for any real-valued function F which is defined and continuous over R, there exists a sequence of polynomials which converges pointwise to F, the convergence being uniform on every bounded subset of R.

6. Prove Theorem 7.7.5 by means of the following steps. Suppose, if possible, that there exists c in I such that $f(c) > 0$. Define $g:I \to \mathsf{R}$ by $g(x) = \max\{0, f(x)\}$ for all x in I. Show that $g \in \mathscr{C}(I)$ and that

$$\int_a^b f(x)g(x)\,dx > 0.$$

Deduce that there exists a polynomial p such that

$$\int_a^b f(x)\{g(x)-p(x)\}\,dx < \int_a^b f(x)g(x)\,dx.$$

Hence obtain a contradiction.

7.8 Stone's generalization of the Weierstrass theorem

Let σ_0 denote the usual supremum metric associated with $\mathscr{C}(I)$; then Weierstrass' theorem asserts that the set P_0 of all real polynomials is dense in $(\mathscr{C}(I), \sigma_0)$. In this section we prove an extremely important generalization due to Stone (this result is usually called the Stone-Weierstrass theorem). Its importance arises from its many applications in abstract analysis; these fall well outside the scope of the present text, but the reader is referred to Stone (1962). There Stone describes his result in a very general setting and gives various applications. There are many variations of the Stone-Weierstrass theorem; the form given below is one of the simplest of these. We present only the underlying ideas involved in generalizing the Weierstrass theorem in the most basic possible case.

Let (X, ρ) be a compact metric space and let $\mathscr{C}(X)$ denote the set of all functions $f:(X, \rho) \to (\mathsf{R}, \mathsf{d})$ which are continuous. Associate with $\mathscr{C}(X)$ the metric σ defined by

$$\sigma(f, g) = \sup_{x \in X} |f(t)-g(t)|;$$

then (see §5.8) $(\mathscr{C}(X), \sigma)$ is a metric space. Our aim is to find a set P which is everywhere dense in $(\mathscr{C}(X), \sigma)$ and which reduces to the set P_0 of polynomials when $X = I$ and ρ is the Euclidean metric. To do this we note that P_0 has the following properties.

(i) The functions† 1, x are in P_0.

(ii) If $f, g \in P_0$ and $\alpha \in \mathsf{R}$ then $f+g, fg, \alpha f \in P_0$.

(iii) For any pair of distinct elements c, d of I, there exists a function f in P_0 such that $f(c) \neq f(d)$; clearly the function x has this property.

It will now be shown how the generalization of (i)–(iii) leads to an extension of Weierstrass' theorem.

First we define some terms.

DEFINITION 7.8.1. Let X be any non-empty set; any collection A of real-valued functions defined on X such that if $f, g \in A$ and $\alpha \in \mathsf{R}$ then $f+g, fg, \alpha f \in A$, is called an *algebra of functions* on X.

(More generally an algebra is essentially a system having the combined structure of a vector space and a ring.) We give some examples.

(i) Let X be any non-empty set; then the set of all real-valued functions on X is an algebra.

(ii) Let (X, ρ) be a metric space; then the set of all real-valued continuous functions on (X, ρ) is an algebra.

† More precisely we mean the constant function whose value is 1 for all x in I and the identity function i.

(iii) Let X be an open interval of R; then the set of all real-valued functions differentiable on X is an algebra.

(iv) The set of all real-valued polynomials on I is an algebra.

DEFINITION 7.8.2. Let (X, ρ) be a metric space and let $A \subseteq \mathscr{C}(X)$ be an algebra of functions on X. Then A is said *to separate points* of X if, for every pair of distinct elements x, y of X, there exists f in A such that $f(x) \neq f(y)$.

Thus the set P_0 of all polynomials separates points of any interval I since P_0 contains the identity function. An example of an algebra which does not separate points of the underlying set is the set of all polynomials of even degree over $[-1, 1]$; for $f(-x) = f(x)$ for all x in $[-1, 1]$ and all such polynomials f.

LEMMA 7.8.1. *If (X, ρ) is a compact metric space and $A \subseteq \mathscr{C}(X)$ is an algebra, then $\bar{A}$ is also an algebra in $\mathscr{C}(X)$, where closure is relative to the supremum metric σ defined above.*

Proof. This is left as an exercise.

We can now state and prove the main result.

THEOREM 7.8.1. *Let (X, ρ) be a compact metric space and $(\mathscr{C}(X), \sigma)$ be the metric space defined above. Then any subset A of $\mathscr{C}(X)$ such that*

(i) *the constant function whose value is* 1 *is in A,*

(ii) *A is an algebra of functions on X,*

(iii) *A separates the points of X,*

is everywhere dense in $(\mathscr{C}(X), \sigma)$.

Proof. This is divided into stages.

I. *To show that if $f \in \bar{A}$ then $|f| \in \bar{A}$.*

First observe that if $f \in A$ then $f \in \mathscr{C}(X)$ and hence $|f| \in \mathscr{C}(X)$.

Since (X, ρ) is compact and $f \in \mathscr{C}(X)$ it follows that $f(X)$ is a bounded subset of R; let $M = \sup_x |f(x)|$. Let $\varepsilon > 0$ be given. Then by Lemma 7.7.2 there exists a polynomial p such that

$$\big||t| - p(t)\big| < \varepsilon \tag{7.8.1}$$

for all t in $[-M, M]$; set

$$p(t) = \sum_{i=0}^{n} c_i t^i.$$

Since $f \in A$ it follows that $p \circ f$, that is the function defined by

$$(p \circ f)(x) = p\{f(x)\} = \sum_{i=0}^{n} c_i\{f(x)\}^i,$$

is also in A. Moreover (7.8.1) implies that

$$\big||f(x)|-(p \circ f)(x)\big| < \varepsilon$$

for all x in X; thus $\sigma(|f|, p \circ f) < \varepsilon$. Hence $|f| \in \bar{\bar{A}} = \bar{A}$.

It will be recalled that when introducing Weierstrass' theorem in §7.7 it was explained that a proof would be given which would extend, in part, to the more general case being considered in the present section; in this connection the reader should note how Lemma 7.7.2 has been used in the first stage of the present proof.

II. *To show that if* $f, g \in \bar{A}$ *then* $\max(f, g)$, $\min(f, g)$ *are also in* $\bar{A}$.

Using the identity

$$\max(f, g)(x) = \tfrac{1}{2}\{f(x)+g(x)+|f(x)-g(x)|\},$$

it follows that if $f, g \in \bar{A}$ then $\max(f, g)$ is in $\bar{A}$; similarly for $\min(f, g)$.

Likewise if $f_1, \dots, f_n \in \bar{A}$, then $\max(f_1, \dots, f_n)$, $\min(f_1, \dots, f_n)$ are also in $\bar{A}$.

III. *To show that if* x_1, x_2 *are distinct elements of* X *and* $\alpha_1, \alpha_2 \in \mathsf{R}$ *then there exists a function* h *in* A *such that* $h(x_1) = \alpha_1$, $h(x_2) = \alpha_2$.

By hypothesis (iii), there exists ϕ in A such that $\phi(x_1) \neq \phi(x_2)$; let h be defined by

$$h(t) = \alpha_1+(\alpha_2-\alpha_1)\frac{\phi(t)-\phi(x_1)}{\phi(x_2)-\phi(x_1)}.$$

Then the function h clearly satisfies the required conditions.

IV. *To show that, given any* f *in* $\mathscr{C}(X)$, *any* x *in* X, *and any* $\varepsilon > 0$, *there exists a function* g_x *in* $\bar{A}$ *such that* $g_x(x) = f(x)$ *and*

$$g_x(t) > f(t)-\varepsilon \tag{7.8.2}$$

for all t *in* X.

To every y in X, by applying III to the elements $f(x), f(y)$ in R it follows that there corresponds a function h_y in A such that

$$h_y(x) = f(x), \quad h_y(y) = f(y).$$

Since $h_y(y)-f(y) = 0$ and since the function h_y-f is continuous over X, there exists an open sphere $S(y, \delta_y)$ of X such that $|h_y(t)-f(t)| < \varepsilon$ for all t in $S(y, \delta_y)$, and so

$$h_y(t) > f(t)-\varepsilon \tag{7.8.3}$$

for all t in $S(y, \delta_y)$.

The collection $\{S(y, \delta_y):y \in X\}$ is an open covering of (X, ρ); since the latter is compact there exists a finite subcovering, say

$$\{S(y_i, \delta_{y_i}):i = 1, ..., n\},$$

by the spheres $S(y, \delta_y)$. Let $g_x = \max(h_{y_1}, ..., h_{y_n})$; then by II, $g_x \in \bar{A}$ and from (7.8.3) it follows that $g_x(t) > f(t)-\varepsilon$ for all t in X. Clearly $g_x(x) = f(x)$.

V. *To obtain the final conclusion.*

We repeat the argument of IV.

Since $g_x(x) = f(x)$, by the continuity of g_x-f there exists an open sphere $S(x, \delta_x)$ such that $|g_x(t)-f(t)| < \varepsilon$ for all t in $S(x, \delta_x)$ and hence

$$g_x(t) < f(t)+\varepsilon \tag{7.8.4}$$

for all t in $S(x, \delta_x)$. Then the collection $\{S(x, \delta_x):x \in X\}$ is an open covering of (X, ρ); again by the compactness of the latter there exists a finite subcovering

$$\{S(x_i, \delta_{x_i}):i = 1, ..., m\},$$

say. Let $F = \min(g_{x_1}, ..., g_{x_m})$, so $F \in \bar{A}$; then from (7.8.4), $F(t) < f(t)+\varepsilon$ for all t in X, and from (7.8.2), $F(t) > f(t)-\varepsilon$ for all t in X. Hence $\sigma(F,f) < \varepsilon$.

It follows that $f \in \bar{A}$ and therefore $\bar{A} = \mathscr{C}(X)$.

EXERCISES 7.8

1. In the notation of Theorem 7.8.1, show that $f \in A$ does not imply that $|f| \in A$.

2. Let $X = [0, 2\pi]$; if $f: X \to \mathsf{R}$ is such that $f(0) = f(2\pi)$ and is continuous over X show that, given any $\varepsilon > 0$ there exists a trigonometric polynomial p, that is a function of the form

$$p(\theta) = a_0 + \sum_{k=1}^{n} (a_k \cos k\theta + b_k \sin k\theta),$$

such that $|f(\theta)-p(\theta)| < \varepsilon$ for all θ in $[0, 2\pi]$.

3. Let X be a compact subset of R^p and let $f\,(= (f_1, \ldots, f_q)) : X \to \mathsf{R}^q$ be continuous (the metrics to be the relevant Euclidean ones). Given any $\varepsilon > 0$, show that there exists a function $g\,(= (g_1, \ldots, g_q)) : X \to \mathsf{R}^q$ of the form

$$g_i(x) = \sum a^{(i)}_{k_1 \ldots k_p} x_1^{k_1} \ldots x_p^{k_p},$$

the sum being taken over a finite number of p-tuples of indices $(k_1, \ldots, k_p)$ ($k_1, \ldots, k_p$ being non-negative integers) where $x = (x_1, \ldots, x_p)$, and such that

$$\sum_{i=1}^{q} \{f_i(x) - g_i(x)\}^2 < \varepsilon$$

for all x in X.

4. Let (X, ρ), (Y, σ) be two metric spaces, $V = X \times Y$ and τ be the metric on V defined by

$$\tau((x, y), (x', y')) = \max\,(\rho(x, x'), \sigma(y, y'))$$

for all (x, y), (x', y') in V. If $f \in \mathscr{C}(V)$ show that, given any $\varepsilon > 0$, there exists a function $p : V \to \mathsf{R}$ of the form

$$p(x, y) = f_1(x)g_1(y) + \ldots + f_n(x)g_n(y)$$

where $f_i \in \mathscr{C}(X)$, $g_i \in \mathscr{C}(Y)$, $i = 1, \ldots, n$, such that

$$|f(x, y) - p(x, y)| < \varepsilon$$

for all (x, y) in V.

5. Let (X, ρ) be a compact metric space and let Y be a non-empty subset of $\mathscr{C}(X)$; let the intersections of all algebras A such that $Y \subseteq A \subseteq \mathscr{C}(X)$ be denoted by $\mathscr{A}(Y)$. Show that $\mathscr{A}(Y)$ is an algebra (it is called the algebra *generated by* Y).

Show that the algebra generated by $\{1, x^{2n+1}\}$, where $n = 0, 1, 2, \ldots$, is everywhere dense in $(\mathscr{C}(I), \sigma)$, σ being the usual supremum metric.

For each positive integer n, for which closed intervals I is the algebra generated by $\{1, x^n\}$ everywhere dense in $(\mathscr{C}(I), \sigma)$?

D. SEPARABILITY

7.9 Separable metric spaces

In this section we look at metric spaces which contain an everywhere dense subset which is countable; the theory of such spaces is important, in particular, in the study of Hilbert spaces. (A Hilbert space H

is essentially a vector space having an inner product defined on the space; furthermore the inner product defines, in a natural and immediate way, a metric on H which makes H a complete metric space.)

DEFINITION 7.9.1. A metric space (X, ρ) is said to be *separable* if it contains an everywhere dense subset which is countable.

For example (R, d) is separable since Q is countable and everywhere dense in (R, d). Other examples are given in §7.10.

THEOREM 7.9.1. *If (X, ρ), (X', ρ') are homeomorphic, then (X, ρ) is separable if and only if (X', ρ') is separable.*

Proof. Let $f:(X, \rho) \to (X', \rho')$ be a homeomorphism. Suppose (X, ρ) is separable and let Y be a countable everywhere dense subset of (X, ρ). Then $f(Y)$ is countable; moreover by Theorem 3.3.1

$$\overline{f(Y)} = f(\bar{Y}) = f(X) = X',$$

so $f(Y)$ is everywhere dense in (X', ρ'). Thus (X', ρ') is separable.

A slight modification of the preceding proof establishes the following result.

COROLLARY 7.9.1. *If (X, ρ), (X', ρ') are two metric spaces such that (X, ρ) is separable, if $f:(X, \rho) \to (X', \rho')$ is continuous and $Y' = f(X)$, then $(Y', \rho'_{Y'})$ is also separable.*

LEMMA 7.9.1. *Let (X, ρ) be a separable metric space and let $Y \subseteq X$; then (Y, ρ_Y) is separable.*

Proof. Let $A = \{x_k : k = 1, 2, \ldots\}$ be a countable everywhere dense subset of (X, ρ). Then for each x_k in A and each positive integer n, either $S(x_k, n^{-1}) \cap Y$ is empty, or it contains some point y_{k_n}, say; let B denote the set of all such points y_{k_n}, so B is countable. Moreover B is everywhere dense in (Y, ρ_Y). To see this let $y \in Y$ and let $\varepsilon > 0$ be given; choose a positive integer m such that $m^{-1} < \frac{1}{2}\varepsilon$. Then there exists x_k in A such that $\rho(x_k, y) < m^{-1}$; hence

$$S(x_k, m^{-1}) \cap Y \neq \varnothing.$$

Taking the corresponding point y_{k_m} of B it follows that

$$\rho(y, y_{k_m}) \leqslant \rho(y, x_k) + \rho(x_k, y_{k_m}) < \varepsilon.$$

This concludes the proof.

We now give some results concerning the relationship between total boundedness (or compactness) and separability of metric spaces.

THEOREM 7.9.2. *Any totally bounded metric space is separable.*

Proof. Let (X, ρ) be totally bounded. Then for each positive integer n there exists a finite n^{-1}-net, say Y_n; let Y denote the union of all the sets Y_n. It is easily seen that Y is countable and everywhere dense in (X, ρ).

Theorem 7.9.2 implies, of course, that every compact metric space is separable.

On the other hand a separable metric space is not necessarily totally bounded; consider, for example, (R, d). However separable spaces do possess a less strong property concerning open coverings, namely that every open covering of a separable metric space contains a countable subcovering (this assertion is known as Lindelöf's theorem). The converse also holds, that is, if a metric space has the property that every open covering of it contains a countable subcovering then it is separable. Before establishing these assertions we introduce another term which helps us to develop these ideas.

DEFINITION 7.9.2. Let (X, ρ) be a metric space and let $\mathscr{T}$ denote the collection of all open sets of (X, ρ). A collection $\{A_\lambda : \lambda \in \Lambda\}$ of open sets of (X, ρ), that is a subset of $\mathscr{T}$, is called a *base* for $\mathscr{T}$ if, for every open set Y, there exists $\Lambda' \subseteq \Lambda$ such that

$$Y = \bigcup_{\lambda \in \Lambda'} A_\lambda.$$

LEMMA 7.9.2. *The collection* $\{S(x, \varepsilon) : x \in X, \varepsilon > 0\}$ *of all open spheres of any metric space* (X, ρ) *is a base for the collection of all open sets of* (X, ρ).

Proof. This is left as a simple exercise.

THEOREM 7.9.3. *Let* (X, ρ) *be a metric space; then the following statements are equivalent:*

(i) (X, ρ) *is separable;*

(ii) *every open covering of* (X, ρ) *contains a countable subcovering;*

(iii) *there exists a countable base for the open sets of* (X, ρ).

Proof. Assume (i). Let $Y = \{y_n : n = 1, 2, \ldots\}$ be a countable everywhere dense subset of (X, ρ); let

$$Z = \{S(y_n, k^{-1}) : n, k = 1, 2, \ldots\},$$

so that Z is a countable collection of open spheres. If x is any point of an open subset G of (X, ρ) there exist integers n, k such that

$$x \in S(y_n, k^{-1}) \subseteq G.$$

For, let $\delta > 0$ be such that $S(x, \delta) \subseteq G$, and let k be a positive integer such that $k^{-1} < \frac{1}{2}\delta$; then there exists y_n in Y such that $\rho(x, y_n) < k^{-1}$, and hence

$$x \in S(y_n, k^{-1}) \subseteq S(x, \delta) \subseteq G.$$

Now let $\{A_\lambda : \lambda \in \Lambda\}$ be any open covering of (X, ρ). If x is a point in one of the sets A_λ then there exist integers n_x, k_x such that

$$x \in S(y_{n_x}, k_x^{-1}) \subseteq A_\lambda.$$

Let $\mathscr{S}$ denote the collection of all those open spheres $S(y_n, k^{-1})$ which are contained in at least one of the sets A_λ. It is seen that the collection $\mathscr{S}$ of such open spheres covers (X, ρ), and that $\mathscr{S}$ is countable. With each sphere S in $\mathscr{S}$ associate exactly one of the sets A_λ containing it; hence the collection of all such A_λ covers (X, ρ) and is countable. This establishes (ii).

Assume (ii); then (iii) follows immediately.

Assume (iii). Let $\{A_n : n = 1, 2, \ldots\}$ be a countable base of the open sets of (X, ρ) and assume that $A_n \neq \varnothing$ for all n. For each n choose an element a_n of A_n and let $Y = \{a_n : n = 1, 2, \ldots\}$. It will be shown that $\bar{Y} = X$, so (X, ρ) is separable.

For let $x \in X$ and let $\varepsilon > 0$ be given. Then $S(x, \varepsilon)$ can be expressed as the union of some of the sets A_n. Hence there exists at least one set, say A_{n_0}, such that $A_{n_0} \subseteq S(x, \varepsilon)$, and so $\rho(x, a_{n_0}) < \varepsilon$; thus $\bar{Y} = X$. This concludes the proof.

Exercises 7.9

1. Show that the interiors of all the open rectangles in R^2, with sides parallel to the coordinate axes of R^2, form a base for the open sets of $(\mathsf{R}^2, \mathsf{d})$. Describe some other bases for the open sets of $(\mathsf{R}^2, \mathsf{d})$.

2. If a metric space (X, ρ) is the union of a countable family of separable subspaces, show that (X, ρ) is separable.

3. Show that the metric spaces (X, ρ), (X, ρ') defined in Exercise 2.2.2 are separable if and only if the spaces (X_i, ρ_i), $i = 1, \ldots, m$ are all separable.

4. Let Y be a subset of a separable metric space (X, ρ); using the equivalent characterization (iii) of Theorem 7.9.3 for separability, show that (Y, ρ_Y) is separable.

5. In the notation of Exercise 2.2.2, show that $\{Y_{\lambda_i}: \lambda_i \in \Lambda_i\}$ is a base for the open sets of (X_i, ρ_i) for $i = 1, \ldots, m$ if and only if

$$\{Y_{\lambda_1} \times \ldots \times Y_{\lambda_m}: \lambda_i \in \Lambda_i, i = 1, \ldots, m\}$$

is a base for the open sets of (X, ρ). Hence give an alternative solution to Exercise 7.9.3.

6. Let (X, ρ) be a separable metric space.

Let $\{A_\lambda: \lambda \in \Lambda\}$ be an open covering of (X, ρ) such that distinct sets A_λ, $A_{\lambda'}$ are disjoint. Prove that Λ is countable.

If Y is a subset of (X, ρ) such that every point of Y is an isolated point, then show that Y is countable. Deduce that the set of all isolated points of (X, ρ) is countable.

Also deduce that any uncountable subset Z of (X, ρ) must contain a limit point of Z.

7. [For this exercise the reader will need to know a little more about cardinality than is summarized in §1.4.]

Let (X, ρ) be a separable metric space. Prove that the set of all open sets of (X, ρ) has cardinality not greater than the cardinality of R.

Deduce that X also has cardinality not greater than that of R.

7.10 Examples concerning the separability of metric spaces

According to our custom let us examine the examples of metric spaces given in §2.2 to see whether or not they are separable.

(i), (ii) (R, d) is separable, as we have already observed. Likewise $(\mathsf{R}^m, \mathsf{d})$ is separable, since Q^m is countable and everywhere dense in $(\mathsf{R}^m, \mathsf{d})$. See also Exercise 7.9.3.

(iii), (iv) $X = \mathsf{R}^m$,

$$\rho(x, y) = \{\textstyle\sum_i |x_i - y_i|^p\}^{1/p},$$

where $p \geqslant 1$, and

$$\rho'(x, y) = \max_i |x_i - y_i|.$$

Since the metrics ρ, ρ' are equivalent to each other and to the Euclidean metric on R^m, by Theorem 7.9.1 it follows that (R^m, ρ), (R^m, ρ') are both separable.

(v) Let X be any non-empty set and ρ be the standard discrete metric. Since every subset of (X, ρ) is closed the only everywhere dense subset is X itself. Therefore (X, ρ) is separable if and only if X is countable.

In view of (i) of Exercise 3.4.6, any discrete space is separable if and only if it is countable.

(vi) $X = \ell^p$ and

$$\rho(x, y) = \{\textstyle\sum_i |x_i - y_i|^p\}^{1/p},$$

where $p \geqslant 1$. Then (ℓ^p, ρ) is separable.

Let Y denote the set of all elements y of ℓ^p such that only a finite number of the coordinates y_i of y are non-zero, and the non-zero coordinates are all rational.

First it is established that Y is countable. Let Y_r denote the subset of Y consisting of all those elements $y\,(= (y_i))$ such that $y_i = 0$ for all $i > r+1$; then

$$Y = \bigcup_{r=1}^{\infty} Y_r.$$

Since $Y_1, Y_2, \ldots$ are all countable, Y is countable.

Next it is shown that Y is everywhere dense. Let $x \in \ell^p$, $y \in Y$ and set

$$y = \{y_1, \ldots, y_r, 0, 0, \ldots\};$$

then

$$\{\rho(x, y)\}^p = \sum_{i=1}^{r} |x_i - y_i|^p + \sum_{i=r+1}^{\infty} |x_i|^p.$$

Given any $\varepsilon > 0$, since $x \in \ell^p$, there exists r such that

$$\sum_{i=r+1}^{\infty} |x_i|^p < \tfrac{1}{2}\varepsilon^p;$$

fix r. Choose $y_1, \ldots, y_r$ in Q such that

$$|x_i - y_i| < \varepsilon/(2r)^{1/p};$$

this is possible since Q is everywhere dense in (R, d). Hence $\rho(x, y) < \varepsilon$ where $y \in Y$, so Y is everywhere dense in (ℓ^p, ρ). Thce (ℓ^p, ρ) is separable.

It was mentioned at the beginning of §7.9 that separability is a concept of some importance in the study of Hilbert spaces; in connection with this it turns out that any Hilbert space H which is separable (that is a space H which, with the metric arising from the inner product of H, is a separable metric space) is isometric to (ℓ^2, ρ) and moreover the vector space structure of H is isomorphic to the vector space structure of ℓ^2. Thus (ℓ^2, ρ) is an exact copy, analytically and algebraically, of any separable Hilbert space. This makes (ℓ^2, ρ) an important example.

(vii) $X = m$ and $\rho(x, y) = \sup_i |x_i - y_i|$. Then (m, ρ) is not separable.

Let Y be the subset of m consisting of all sequences whose elements are 0 or 1. Let $\mathscr{S}$ denote the collection $\{S(y, \frac{1}{2}) : y \in Y\}$ of open spheres; then the spheres $S(y, \frac{1}{2})$, $S(y', \frac{1}{2})$ in $\mathscr{S}$ are disjoint provided $y \neq y'$. Suppose that Z is everywhere dense in (m, ρ); then each of the spheres $S(y, \frac{1}{2})$ would contain at least one point of Z. It will be shown that Y, and hence the collection $\mathscr{S}$, is not countable; thus Z is not countable, and so (m, ρ) is not separable.

To show that Y is not countable we use a standard contradiction argument. Thus suppose, if possible, that we can express Y as

$$\{y_n : n = 1, 2, \ldots\}.$$

Let $y_n = (y_{n1}, y_{n2}, \ldots)$ where each y_{nm} is either 0 or 1. Define $z = (z_1, z_2, \ldots)$ as follows; each z_m is to be either 0 or 1 (so $z \in Y$) and such that $z_1 \neq y_{11}$, $z_2 \neq y_{22}$, and so on. This implies that $z \neq y_1$ (since z and y_1 differ in their first coordinates), $z \neq y_2$ (since z and y_2 differ in their second coordinates), and so on; hence z is not equal to any of the elements y_n, so $z \notin Y$, which is a contradiction.

(viii) $X = \mathscr{C}(I)$ and ρ is the usual supremum metric. Then $(\mathscr{C}(I), \rho)$ is separable.

To see this we first observe that by the Weierstrass approximation theorem the set of all polynomials is everywhere dense in $(\mathscr{C}(I), \rho)$; however this alone is not sufficient to establish the separability of $(\mathscr{C}(I), \rho)$ since the set of all polynomials is not countable (for example consider the subset consisting only of constants). Instead we consider the set P^* of polynomials with rational coefficients; it will be shown that P^* is countable and everywhere dense in $(\mathscr{C}(I), \rho)$.

Let P_n^* be the subset of P^* consisting of all polynomials, with rational coefficients, of degree not greater than n; then there exists a bijection between P_n^* and $\mathbb{Q}^{n+1}$, so P_n^* is countable. But

$$P^* = \bigcup_{n=0}^{\infty} P_n^*,$$

so P^* is countable.

Let $f \in \mathscr{C}(I)$ and $\varepsilon > 0$ be given; let p be a polynomial such that $\rho(f, p) < \frac{1}{2}\varepsilon$ and let

$$p(t) = a_0 + a_1 t + \ldots + a_n t^n$$

where $a_i \in \mathbb{R}$ for $i = 0, 1, \ldots, n$. Let $c = \max(|a|, |b|)$; for each i $(0 \leqslant i \leqslant n)$ choose b_i in $\mathbb{Q}$ such that

$$|b_i - a_i| < \frac{\varepsilon}{2(n+1)c^i},$$

and let
$$q(t) = b_0 + b_1 t + \ldots + b_n t^n.$$

Then
$$|p(t) - q(t)| = \left|\sum_{i=0}^{n} (b_i - a_i)t^i\right| < \tfrac{1}{2}\varepsilon,$$

so $\rho(p, q) \leqslant \frac{1}{2}\varepsilon$. Hence $\rho(f, q) < \varepsilon$, so P^* is dense in $(\mathscr{C}(I), \rho)$.

(ix) $X = \mathscr{B}(I)$ and ρ is the usual supremum metric. We can no longer use the Weierstrass approximation theorem since the functions which we are considering are not necessarily continuous over I. In fact $(\mathscr{B}(I), \rho)$ is not separable; we outline the proof, leaving the details to the reader.

Let Y be the subset of $\mathscr{B}(I)$ consisting of all the characteristic functions, that is the functions defined on I which take only the values 0 or 1; this set is not countable. Let Z be a set which is everywhere dense in $(\mathscr{B}(I), \rho)$; by using a similar argument to that employed in (vii) it will follow that Z is not countable.

(x) $X = \mathscr{C}(I)$ and

$$\rho(f, g) = \int_a^b |f(t) - g(t)|\, dt.$$

Let P^* be defined as in (viii) and let $f \in \mathscr{C}(I)$; then given any $\varepsilon > 0$ there exists q in P^* such that

$$\sup_t |f(t) - q(t)| < \frac{\varepsilon}{b-a}$$

so
$$|f(t) - q(t)| < \frac{\varepsilon}{b-a}$$

for all t in I, and hence

$$\int_a^b |f(t)-q(t)|\,dt < \varepsilon;$$

thus $\rho(f, q) < \varepsilon$. Hence P^* is everywhere dense in $(\mathscr{C}(I), \rho)$ which is therefore separable.

EXERCISES 7.10

1. Show that the metric space defined in Exercise 2.2.3 is separable.

2. Are the metric spaces defined in Exercises 2.1.5, 2.1.6 separable?

3. Let X denote the set of all real valued functions which are defined, continuous, and bounded over R; let $\rho: X \times X \to \mathsf{R}$ be defined by

$$\rho(x, y) = \sup_{-\infty < t < \infty} |x(t)-y(t)|.$$

Is (X, ρ) separable?

APPENDIX

There are a number of properties and results concerning metric spaces which have been established and which can usefully be collected together for the sake of comparison; this is now done.

A.1 Properties of metric spaces

First it is convenient to introduce a non-standard term.

DEFINITION A.1.1. Let (X, ρ), (X', ρ') be two metric spaces and $f:(X, \rho) \to (X', \rho')$ be a bijection of X onto X' such that f is uniformly continuous over X and f^{-1} is uniformly continuous over X'; then f is said to be a *uniform homeomorphism* and (X, ρ), (X', ρ') are said to be *uniformly homeomorphic*.

THEOREM A.1.1. *Let* (X, ρ), (X', ρ') *be homeomorphic. Then* (X, ρ) *is discrete, or compact, or connected, or separable, if and only if* (X', ρ') *has the same property.*

See (i) of Exercise 3.3.2 and §§5.2, 6.1, 7.9.

On the other hand we cannot add the properties of being complete or totally bounded to the above list; for example the spaces $((0, 1), \mathsf{d})$, (R, d) are homeomorphic but only one is complete and only one is totally bounded. However we do have the following result.

THEOREM A.1.2. *Let* (X, ρ), (X', ρ') *be uniformly homeomorphic. Then* (X, ρ) *is complete, or totally bounded, if and only if* (X', ρ') *has the same property.*

See Exercises 4.1.5, 5.8.4.

THEOREM A.1.3. *Let* $f:(X, \rho) \to (X', \rho')$ *be continuous and let* $Y' = f(X)$. *If* (X, ρ) *is compact, or connected, or separable, then* $(Y', \rho'_{Y'})$ *has the same property.*

See §§5.8, 6.4, 7.9.

THEOREM A.1.4. *Let* $f:(X, \rho) \to (X', \rho')$ *be uniformly continuous and let* $Y' = f(X)$. *If* (X, ρ) *is totally bounded then* $(Y', \rho'_{Y'})$ *is totally bounded.*

See Exercise 5.8.4.

Note however that if (X, ρ) is discrete, or complete, then $(Y', \rho'_{Y'})$ is not necessarily discrete, or complete, respectively. See the example in §3.3, and Exercise 4.1.6.

DEFINITION A.1.2. Let P be any property which is applicable to a metric space; P is called a *hereditary property* if whenever a metric space (X, ρ) has the property P, then any subspace of (X, ρ) also has the property P.

THEOREM A.1.5. *The properties of discreteness, boundedness, total boundedness, separability are all hereditary; the properties of completeness, compactness, connectedness are not hereditary.*

For the assertions concerning total boundedness and separability see Lemma 5.7.3 and Lemma 7.9.1 respectively. All of the negative assertions are easily established by example. Concerning completeness and compactness we do know, however, that a closed subset of a complete (or compact) metric space is complete (or compact, respectively); see Theorem 4.1.1 and Lemma 5.7.4.

A.2 Product spaces

Let (X_i, ρ_i), $i = 1, \dots, m$ be metric spaces, $X = X_1 \times \dots \times X_m$; let $x = (x_1, \dots, x_m)$, $y = (y_1, \dots, y_m)$ where $x_i, y_i \in X_i$, and define

$$\rho(x, y) = \max_{1 \leq i \leq m} \rho_i(x_i, y_i), \quad \rho'(x, y) = \sum_{i=1}^{m} \rho_i(x_i, y_i)$$

and more generally

$$\rho''(x, y) = \left\{ \sum_{i=1}^{m} [\rho_i(x_i, y_i)]^p \right\}^{1/p},$$

where $p \geqslant 1$. Then ρ, ρ' are metrics on X (see Exercise 2.2.2); likewise it can be established that ρ'' is a metric on X.

Clearly

$$\rho(x, y) \leqslant \rho''(x, y) \leqslant m^{1/p} \rho(x, y) \tag{A.2.1}$$

for all x, y in X and all $p \geqslant 1$. Hence we have

THEOREM A.2.1. *In the above notation, the metrics ρ, ρ', ρ'' are uniformly equivalent.*

We shall denote an open sphere in (X_i, ρ_i) by S_i and an open sphere in (X, ρ), (X, ρ'), (X, ρ'') by S, S', S'' respectively; a corresponding convention is used for closed spheres.

The inequalities (A.2.1) imply

$$S''(x, r) \subseteq S(x, r) \subseteq S''(x, m^{1/p}r),$$

and

$$\bar{S}''(x, r) \subseteq \bar{S}(x, r) \subseteq \bar{S}''(x, m^{1/p}r),$$

for any x in X and any $r > 0$.

THEOREM A.2.2. (i) *For any x in X and any $r > 0$,*

$$S(x, r) = S_1(x_1, r) \times \ldots \times S_m(x_m, r)$$

$$\bar{S}(x, r) = \bar{S}_1(x_1, r) \times \ldots \times \bar{S}_m(x_m, r).$$

Analogous identities do not hold for $S''(x, r)$, $\bar{S}''(x, r)$.

(ii) *If $Y_i \subseteq X_i$ for each i, then*

$$(Y_1 \times \ldots \times Y_m)^\circ = Y_1^\circ \times \ldots \times Y_m^\circ$$

$$\overline{Y_1 \times \ldots \times Y_m} = \bar{Y}_1 \times \ldots \times \bar{Y}_m$$

where Y_i° is the interior of Y_i in (X_i, ρ_i) and $(Y_1 \times \ldots \times Y_m)^\circ$ is the interior of $Y_1 \times \ldots \times Y_m$ in (X, ρ) or (X, ρ''), and similar comments apply to the closures.

Moreover if $Y_i \neq \varnothing$ for each i, then $Y_1 \times \ldots \times Y_m$ is open (or closed) in (X, ρ) or (X, ρ''), if and only if each Y_i is open (or closed, respectively) in (X_i, ρ_i).

(iii) *The sequence $(x^{(n)})$ in (X, ρ), or (X, ρ''), is convergent (or fundamental) if and only if each sequence $(x_i^{(n)})$ is convergent (or fundamental) in (X_i, ρ_i). Moreover the sequence $(x^{(n)})$ converges to x in (X, ρ), or (X, ρ''), if and only if each sequence $(x_i^{(n)})$ converges to x_i where $x = (x_1, \ldots, x_m)$.*

If x is a cluster value of $(x^{(n)})$ in (X, ρ), or (X, ρ''), then x_i is a cluster value of $x_i^{(n)}$ in (X_i, ρ_i). The converse of this is false, however.

The proofs of these assertions are left to the reader.

THEOREM A.2.3. *The projection function $\pi_i:(X, \rho) \to (X_i, \rho_i)$ defined by $\pi_i(x) = x_i$, where $x = (x_1, \ldots, x_m)$, is uniformly continuous over X.*

See Exercise 3.2.8.

Hence if (X, ρ) is compact, or totally bounded, or connected, or separable then (by Theorems A.1.3, A.1.4) each (X_i, ρ_i) has the same property. To this list may be added the properties of boundedness, discreteness, and completeness; to see this we need a different argument. Let $(x_1, \ldots, x_m) \in X$ and let

$$\tilde{X}_1 = X_1 \times \{x_2\} \times \ldots \times \{x_m\};$$

then $(\tilde{X}_1, \rho_{\tilde{X}_1})$ is isometric, and so uniformly homeomorphic, to (X_1, ρ_1). By Theorem A.1.5, if (X, ρ) is bounded or discrete then so also is $(\tilde{X}_1, \rho_{\tilde{X}_1})$ and hence (X_1, ρ_1) has the same property. Furthermore $\tilde{X}_1$ is a closed subset of (X, ρ); why? Thus if (X, ρ) is complete, then so also is $(\tilde{X}_1, \rho_{\tilde{X}_1})$ and again (X_1, ρ_1) has the same property.

In the reverse direction if each (X_i, ρ_i) is bounded, or discrete, or complete, or compact, or totally bounded, or connected, or separable then (X, ρ) has the same property. The assertion concerning boundedness follows immediately from the definition of ρ. To prove the assertion concerning discreteness use, for example, the characterization (iii) of Lemma 2.6.2 of a discrete space; the details are left to the reader. For the remainder of the assertions see Exercises 4.2.2, 5.9.1, 6.4.1, 7.9.3.

To sum up, we have

THEOREM A.2.4. *The metric spaces (X, ρ), (X, ρ'') are bounded, or discrete, or complete, or compact, or totally bounded, or connected, or separable, if and only if each (X_i, ρ_i) has the same property.*

BIBLIOGRAPHY

(i) The following list contains the books and journals referred to in the text.

AHLFORS, L. V. 1966. *Complex Analysis*. McGraw–Hill, 2nd Edition.

APOSTOL, T. M. 1957. *Mathematical Analysis*. Addison-Wesley.

BIELECKI, A. 1956. Une remarque sur la méthode de Banach–Cacciopoli–Tikhonov dans la théorie des équations différentielles ordinaires. *Bulletin de l'Académie Polonaise des Sciences, Cl.* III, **4**, 261–264.

COHEN, L. W. AND EHRLICH, G. 1963. *The Structure of the Real Number System*. Van Nostrand.

COPSON, E. T. 1968. *Metric Spaces*. Cambridge University Press.

DIEUDONNÉ, J. 1960. *Foundations of Modern Analysis*. Academic Press.

FERRAR, W. L. 1938. *A Text-book of Convergence*. Oxford University Press.

FRÉCHET, M. 1906. Sur quelques points du calcul fonctionnel, *Rendiconti del Circolo Matematico di Palermo*, **22**, 1–74.

GLEASON, A. M. 1966. *Fundamentals of Abstract Analysis*. Addison-Wesley.

HERSTEIN, I. N. 1964. *Topics in Algebra*. Ginn–Blaisdell.

HILLE, E. 1969. *Lectures on Ordinary Differential Equations*. Addison-Wesley.

KOLMOGOROV, A. N. AND FOMIN, S. V. 1957. *Elements of the Theory of Functions and Functional Analysis*; Vol. 1, *Metric and Normed Spaces*. Graylock Press.

SIMMONS, G. F. 1963. *Introduction to Topology and Modern Analysis*. McGraw–Hill.

STONE, M. H. 1962. A Generalized Weierstrass Approximation Theorem. *Studies in Modern Analysis* (edited by R. C. Buck). Published by The Mathematical Association of America; distributed by Prentice–Hall.

(ii) The above books by Ahlfors, Copson, Dieudonné, Gleason, Kolmogorov and Fomin, and Simmons contain discussions of the properties of metric spaces; the following are some further texts in which metric spaces are treated:

BOAS, R. P. 1960. *A Primer of Real Functions*. Published by The Mathematical Association of America; distributed by Wiley.

BURGESS, D. C. J. 1966. *Analytical Topology*. Van Nostrand.

GOFFMAN, C. AND PEDRICK, G. 1965. *First Course in Functional Analysis*. Prentice–Hall.

KURATOWSKI, K. 1961. *Introduction to Set Theory and Topology*. Pergamon Press.

WHITE, A. J. 1968. *Real Analysis: an Introduction*. Addison-Wesley.

INDEX